T0135745

Computation of invariant measures with dimension reduction methods

Dissertation
zur Erlangung des Doktorgrades
der Fakultät für Mathematik der Universität Bielefeld

vorgelegt von
Jens Kemper

Bielefeld, im Januar 2010

1. Gutachter: Prof. Dr. W.-J. Beyn

2. Gutachter: Prof. Dr. O. Junge

Datum der mündlichen Prüfung: 22. März 2010

Im Zuge der Veröffentlichung wurde die vorliegende Dissertation redaktionell überarbeitet.

Bielefeld, im März 2010

Bibliografische Information der Deutschen Nationalbibliothek

Die Deutsche Nationalbibliothek verzeichnet diese Publikation in der Deutschen Nationalbibliografie; detaillierte bibliografische Daten sind im Internet über http://dnb.d-nb.de abrufbar.

ISBN 978-3-8325-2452-4

Logos Verlag Berlin GmbH
Comeniushof, Gubener Str. 47,
10243 Berlin
Tel.: +49 (0)30 42 85 10 90
Fax: +49 (0)30 42 85 10 92
INTERNET: http://www.logos-verlag.de

Acknowledgements

At this point I would like to express a big thank-you to all the people who helped me in the realization of this thesis.

First of all, I wish to thank my advisor, Prof. Dr. Wolf-Jürgen Beyn, who gave me the opportunity to study this interesting research project. He supported my work during the whole project with great commitment, always shared his remarkable intuition, gave useful advice and showed me new approaches whenever all approaches seemed to fail.

Also, I would like to thank the various current and former members of our research group 'Numerical Analysis of Dynamical Systems' for helping me with advice and simply providing such a pleasant working atmosphere through all the years.

Further, I wish to thank researchers inside and outside the Department of Mathematics in Bielefeld who shared their knowledge and discussed open problems with me. In particular, I would like to mention Prof. Dr. Ludwig Elsner who gave me a deep insight into many topics of numerical linear algebra. I have benefitted a lot from his ideas and knowledge. Also Prof. Dr. Hans Zessin and Dr. Kim Frøyshov, currently at the University of Zürich, helped me with fruitful discussions mainly concerning distance notions of measures.

This thesis was developed during my work for the Collaborative Research Centre 701 'Spectral Structures and Topological Methods in Mathematics'. I thank the CRC for financial and ideal support.

Last but not least, I wish to thank my family for their support throughout all the highs and lows during the last years. In particular, I thank Annette not only for the tremendous effort to get the text readable but also for her deep understanding and patience. This thesis would not have seen the light of day without her.

Contents

Introduction

To understand a dynamical system, the knowledge of the asymptotic behavior of trajectories for arbitrary initial values is important. This long-time behavior is described by limit-sets, attractors, almost invariant sets and invariant measures.

Generally speaking, there are two classical numerical approaches for analyzing the long-time behavior. The first one uses the simulation of a number of trajectories over large time intervals and in this way tries to get an overall picture of the dynamics. The problem of this approach consists in the arbitrariness of the choice of initial values. The question arises whether the chosen initial values lead to a detection of all relevant information. For example, trajectories remain for long time scales in an almost invariant set before leaving it. The dynamical behavior outside an almost invariant set is in general poorly captured by this approach.

The second approach, which we pursue in this thesis, is to make a global statistical analysis of the underlying system. Here, ergodic theory comes into play, cf. [Sin94]. An invariant measure μ of a dynamical system provides fundamental information about its long-time behavior. A key result is the Birkhoff Ergodic Theorem for ergodic measures μ. It states that for μ-almost every initial value u the so-called *time average* along the trajectory starting at u equals the *space average* measured by μ.

In this thesis we focus on high-dimensional *discrete dynamical systems* defined by

$$u_{i+1} = F(u_i), \quad i = 0, 1, 2, \ldots, \tag{1}$$

where F is a diffeomorphism on a compact subset $X \subset \mathbb{R}^N$ with large $N \gg 1$. The main results of this thesis aim at a computational approach to invariant measures in high-dimensional systems given by (1).

Invariant measures are fixed points of the Frobenius-Perron operator $P = P_F$ mapping the space $\mathcal{M}^1(\mathbb{R}^N)$ of Borel probability measures on $\mathbb{R}^N$ onto itself. P is defined by

$$P(\mu)(A) = \mu(F^{-1}(A)) \quad \text{for all } A \in \mathcal{B}(\mathbb{R}^N), \mu \in \mathcal{M}^1(\mathbb{R}^N). \tag{2}$$

A common approach to approximating invariant measures goes back to Ulam in 1960, [Ula60]. By a discretization of the phase space it associates to P a large nonnegative matrix, called Perron-Frobenius matrix. An eigenvector corresponding to the eigenvalue 1 of this matrix defines a discrete measure approximating an invariant measure of the system. In recent years, a subdivision algorithm based on this idea was developed by Dellnitz, Junge and co-workers ([DFJ01], [DJ98], [DJ99]). In this algorithm the state space X is discretized by a finite box covering. The entries of the Perron-Frobenius matrix approximating P are given by transition probabilities from one box of the collection to another. Since X is mapped into itself, the Perron-Frobenius matrix is (column) stochastic. Then by

1

the Perron-Frobenius Theorem, the existence of a nonnegative fixed point representing a proper discrete measure follows. The approach is combined with an efficient storing algorithm for the boxes and an adaptation in the bisection process. For systems of low dimension, this technique is a powerful tool to compute invariant measures. In [DJ02], [DFJ01], Dellnitz, Junge et al. prove convergence of the discrete measures computed by this so-called *Adaptive Invariant Measure (AIM) algorithm* to a Sinai-Ruelle-Bowen (SRB) measure as the diameter of the box collection converges to zero. A detailed discussion of the convergence theory is part of my diploma thesis [Kem02].

Yet in many applications, the state space of the discrete dynamical system is high-dimensional. For instance, consider the finite element discretization of a parabolic partial differential equation (PDE) of the type

$$u_t = \Delta u + f(u), \quad u(t) \in H, \tag{3}$$

where H is a proper Sobolev space H, e.g. $H_0^1(\Omega)$, and boundary and initial conditions are imposed. In Chapter 6 we analyze the Chafee-Infante problem as an example for (3). Typical parabolic problems—as in the Chafee-Infante case—satisfy the following two assumptions: First, the system has an attractor $\mathcal{A} \subset H$ and a corresponding invariant measure μ whose support is a subset of the attractor. Second, the attractor is part of a low-dimensional manifold $\mathcal{M}$. Results of this type are derived in the theory of inertial manifolds, cf. [Tem97].

Under these assumptions the numerical approximation of an invariant measure appears feasible. Nevertheless, the original AIM algorithm suffers from the 'curse of dimension' when used in higher dimensions. Even when the support of the invariant measure is low-dimensional, the subdivision algorithm has to deal with an exponentially increasing number of boxes in the first recursion steps. If the state space is N-dimensional, 2^N boxes are needed to obtain the first discretization in every coordinate. Since no relevant reduction can be expected before this step, the number of boxes quickly exceeds an amount on which computations can be made even though the relevant dynamics is embedded in a low-dimensional manifold.

There are several ways to approach this problem. We briefly discuss three of them.

- One can simplify the computation of invariant measures by using structural properties of the underlying system. Dellnitz, Hessel-von Molo et al. have presented an ansatz using symmetries of the dynamical system ([MHvMD06], [Jun01]). They showed that the symmetries of the system are linked to the symmetries of eigenmeasures of the Frobenius-Perron operator. This observation can in principle be used to derive the eigenmeasures with less computational costs.

- A new ansatz by Junge and Koltai presented in [Jun09] uses so-called *sparse hierarchical grids* which are based on a tensor product construction. Using the Haar basis of $L^2([0,1]^N)$, finite-dimensional approximation spaces (given by a sparse basis) are derived which have the largest benefit-to-cost ratio. By this approach, a comparable accuracy of the approximation is achieved with a computational effort of considerably lower order than is needed by the standard Ulam basis.

- Another approach, which we follow here, is based on model reduction methods. We focus on the so-called *proper orthogonal decomposition (POD)*. It is described, for instance, in [HLB96]. There are many promising results when using this method for the computation of single trajectories mainly in control theory (see [KV01], [KV02]). For linear systems they can also be combined with a balanced truncation ansatz ([Ant05], [ASG01], [RCM04]). In this thesis, we will use the POD method to derive new algorithms approximating the global dynamics of a high-dimensional system.

The outline of the thesis is as follows. In Chapter 1, we begin with a brief review of the standard theory of dynamical systems with a focus on their long-time behavior. Then the connection between invariant sets and statistical properties arising from ergodic theory is elaborated. This gives us a deeper insight into Ulam's approach which we present in detail. We formulate a version of the AIM algorithm based on this approach at the end of Chapter 1. The algorithm is used as a basis of the reduced-space algorithms that we propose in Chapter 3.

In Chapter 2, we present in detail the POD method we use for model reduction. It relies on the singular value decomposition of linear operators on Hilbert spaces. We define a linear operator Y representing given snapshots and discuss the fundamental result stating that the POD basis of rank ℓ is given by the first ℓ singular vectors of Y. The approximation error is given by the sum of the remaining singular values σ_k, $k > \ell$. Based on this result, Kunisch and Volkwein [KV01] derive error results for the POD method for finite element discretizations of parabolic problems. These error bounds hold for single trajectories on finite time intervals.

With our new POD-based algorithms we aim at the approximation of invariant measures describing the long-time behavior of dynamical systems. We derive explicit error bounds concerning the long-time behavior of POD solutions in some model cases to show that the POD method is appropriate for the algorithms. The proofs are based on perturbation theory for singular decompositions, cf. [Ste90], [GvL96]. We illustrate the theoretical results by several numerical examples. In particular, a linear parabolic equation is studied in some detail. This analysis gives a first insight into our numerical experiments with a nonlinear parabolic equation in Chapter 6.

At the end of Chapter 2, we study the case where more than one trajectory is used for the derivation of snapshots. These considerations prepare the introduction of the POD-based algorithms in Chapter 3 that are based on the computation of many short-time trajectories instead of one long-time trajectory.

After this preparatory work, we formulate our new algorithms in Chapter 3 combining the subdivision technique of the AIM algorithm with the model reduction approach using the POD method. The resulting PODAIM algorithm works in a fixed POD space. We extend it to the algorithm PODADAPT which adaptively changes the POD space during the box refinement process. When the POD space is being changed in the second algorithm, the box collection needs to be transferred. We derive a computational realization for this transformation.

Finally, we give a proper explicit representation for the measures computed by the reduced-space algorithms in preparation of the error analysis presented in the following chapters.

In Chapter 4, we derive a proper distance notion for the various probability measures occurring in the algorithms. The AIM algorithm computes so-called *discrete measures* or *box measures* that are absolutely continuous and have constant densities on every box of the box collection. The algorithms presented in Chapter 3 compute discrete measures with support in a reduced space which are then extended to measures with thin support in the original space.

The usual formulation of the weak metric on $\mathcal{M}^1(X)$, $X \subset \mathbb{R}^N$ compact, is based on integrals of a dense subset of $C(X)$. This formulation implies a numerical realization which does not yield a suitable distance notion for discrete measures. An equivalent metric is given by the Prohorov metric, cf. [Dud02]. This metric takes into account the geometry of the box measures. We derive a proper numerical approximation of the Prohorov metric for given discrete measures with thin support. Then we compare its realization with another possible distance notion using negative Sobolev norms. It turns out that the numerical realization of the Prohorov metric is more robust and efficient than a discretization of the negative Sobolev norm.

At the end of Chapter 4, we show contraction results for the Frobenius-Perron operator on discrete measures in terms of the Prohorov distance. We derive a proper condition under which the Perron-Frobenius matrix is contracting with respect to the discretized Prohorov metric. The sharpness of the condition is shown by low-dimensional examples. It is the underlying idea that these contraction properties lead to convergence of the box measures to an invariant measure of the system with decreasing diameter of the box collection.

In Chapter 5, we continue the analysis of the algorithms developed in Chapter 3. First, we state how existing results concerning subdivision algorithms can be applied in our context. In the original space, the convergence analysis of Dellnitz and Junge mentioned above applies. In principle these results also cover the reduced-space system. Additionally, it is a realistic scenario that the reduced system satisfies the assumptions of the ergodic theory for expanding maps developed by Keller, Blank, Liverani, Baladi et al., cf. [Liv03], [Bal00], [BKL02].

Two results concerning the relation of the approximation processes in the original and the reduced space are given: First, we show that invariant Borel measures in the reduced space extend to invariant measures of the original system if the POD subspace is F-invariant. In particular, every F-invariant measure is of that shape, if the POD subspace is uniformly attracting. Secondly, we show as the main result of this chapter relations between the box measures of the AIM and the PODAIM algorithm. Under structural assumptions on the box collections, we derive an explicit error bound. The proof is based on an entry-wise analysis of the Perron-Frobenius matrices in both spaces. We use perturbation theory for Perron roots developed by Funderlic, Meyer and co-authors in [FM86], [Mey80], [Mey94], based on properties of the group inverse of the Perron-Frobenius matrix. We end the chapter with a slight generalization of results to the case where the POD space is only F-invariant in an approximate sense.

We conclude our thesis with various numerical computations in Chapter 6. Some of the results presented here have already been worked out in the preprint [Kem08]. For a proper visualization of our numerical experiments it is necessary to derive a suitable representation of discrete measures embedded in a high-dimensional space. We suggest a marginal-like representation that is appropriate for spatial discretizations of scalar parabolic problems.

As a first test we define a 10-dimensional system built up from a slightly perturbed embedding of the well-known Lorenz equations. It turns out that the POD approximation used in the PODAIM algorithm detects the embedding with high accuracy as expected. We use this example to analyze the POD computation in detail and relate our observations to the corresponding results in Chapter 2. The result is a trade-off between the length and the number of trajectories used as snapshots.

Finally, we analyze different finite element discretizations of the Chafee-Infante problem. This parabolic system is a good model example since it is well-analyzed and the global attractor is known to be low-dimensional. Due to symmetries in the system the Perron-Frobenius matrix is reducible, i.e. the Perron eigenspace is multi-dimensional. To derive a discrete measure representing the full dynamics of the system, we have to select an appropriate fixed point in the Perron eigenspace. We derive two approaches. One of them is based on the group inverse of the Perron-Frobenius matrix which was used in Chapter 5.

It turns out that the PODAIM algorithm is robust under refinements of the spatial discretization while the AIM algorithm indeed suffers from the 'curse of dimension'. The Prohorov metric is used to compare the results of the algorithms. We discuss in detail the effect of the POD space dimension on the distance between the discrete measures in the PODAIM and the AIM algorithm. Note that the corresponding error bound derived in Chapter 5 depends strongly on this dimension.

Finally we show that the PODADAPT algorithm also gives promising results for this model problem. Nevertheless, the additional computational effort for the adaptation of the POD space does not lead to better approximation results. It remains as an open task to find applications where the approach of the PODADAPT algorithm leads to an additional benefit.

Summarizing, the main results of our thesis are the following: Subdivision algorithms for the approximation of invariant measures in high-dimensional systems are developed. Explicit error bounds concerning the long-time behavior of the POD method used in these algorithms are derived. We propose a discrete version of the Prohorov metric as a proper distance notion for the discrete measures computed by the algorithms. By an error analysis for discrete measures, the POD-based algorithms are compared with the subdivision algorithms developed by Dellnitz, Junge and co-workers. A marginal-like representation for discrete measures in high-dimensional systems is derived and the power of the algorithms is illustrated by various numerical experiments.

Chapter 1

The Adaptive Invariant Measure algorithm

At the beginning of this thesis we give an insight into the theories that are essential for the understanding of the main results. The subject of this thesis is covered by the theory of dynamical systems on the one hand and ergodic theory on the other hand. In Sections 1.1 and 1.2, we introduce the main terms and definitions together with the standard results in these two fields of research. Readers familiar with the theories should omit the first two sections and begin with Ulam's approach in Section 1.3.

Ulam's approach provides the essential idea of the subdivision algorithms presented in this thesis. We define the Frobenius-Perron operator in this section together with corresponding finite-dimensional approximations by Perron-Frobenius matrices resulting from a discretization of the phase space.

The AIM algorithm developed by Dellnitz, Junge and co-workers is presented in the following sections. Our new algorithms presented in Chapter 3 are further developments of this subdivision algorithm and will be compared with the AIM algorithm throughout the thesis.

1.1 Dynamical systems

In this section, the main concepts of the theory of dynamical systems are introduced. We restrict ourselves to the case of a finite dimensional state space. Proofs of the properties in the following remarks can be found in [SH96].

We distinguish between two types of dynamical systems: maps and ordinary differential equations.

Definition 1.1. Given a map $F : X \subset \mathbb{R}^N \to \mathbb{R}^N$, a *discrete dynamical system* on X is defined by the iteration

$$u_{i+1} = F(u_i), \quad u_0 \in X \tag{1.1}$$

if the sequence $(u_i)_i$ stays in X.

Given a map $f \in C(\mathbb{R}^N, \mathbb{R}^N)$ we define a *continous-time dynamical system* via the ordinary differential equation

$$
\begin{aligned}
u_t &= f(u) \\
u(0) &= u_0 \in X \subset \mathbb{R}^N
\end{aligned}
\tag{1.2}
$$

The system (1.2) defines a dynamical system on $X \subset \mathbb{R}^N$ if for every $u_0 \in X$ there is a unique solution $u(t)$ to (1.2) with $u(t) \in X$ for $t \geq 0$.

Remark. *In the subdivision algorithm we work with discrete dynamical systems. Nevertheless we sum up the theory for both types of dynamical systems since it can be formulated in an abstract setting including both types (see Definition 1.2).*

Discrete dynamical systems evolve naturally from continuous-time dynamical systems based on differential equations, e.g. by a time discretization or a Poincaré map.

Connected to every dynamical system there is the evolution operator defining a semigroup.

Definition 1.2. For a discrete dynamical system given by (1.1) we define the *evolution operator* $S : \mathbb{N} \times X \to X$ by $S^n u = F^n(u)$.

For a dynamical system given by (1.2) with solution $u(t) = u(t; t_0, u_0)$ we define the evolution operator $S : \mathbb{R}_+ \times X \to X$ by

$$S(t)u_0 = u(t; t_0, u_0).$$

.

Remark.   *1. Note that the evolution operator defines a semigroup in both cases meaning that the following to properties hold for the operator $S : \mathbb{T} \times X \to X$, $\mathbb{T} \in \{\mathbb{N}, \mathbb{R}_+\}$:*

- $S^0 = Id_E$
- $S^{t+s}(u) = S^{s+t}(u) = S^t(S^s(u)), \quad t, s \in \mathbb{T}$

where Id_E denotes the identity map on E.

2. Many properties of dynamical systems can be formulated via the evolution operator. That means that we do not have to distinguish between discrete and continuous dynamical systems for introducing more properties in the following. We will denote a dynamical system by the semigroup $S : \mathbb{T} \times X \to X$ as in the first remark if we do not want to distinguish between the two types.

Invariant sets and attractors

A central property of dynamical systems is the existence of invariant sets.

Definition 1.3. A set $B \subset X$ is called *positive invariant* for the dynamical system (1.1) and (1.2), respectively, if the semigroup S satisfies

$$S^t B \subset B \quad \text{for all } t \geq 0.$$

The set B is called *negative invariant* if

$$B \subset S^t B \quad \text{for all } t \geq 0.$$

A positive and negative invariant set is called *invariant*.

Remark. *For discrete dynamical systems the conditions for invariance simplify to $F(B) \subset B$ and $B \subset F(B)$.*

A simple example is given by a fixed point *$\bar{u} \in X$ satisfying $S^t(\bar{u}) = \bar{u}$ for $t \geq 0$, also called* equilibrium point *for a continuous-time dynamical system.*

Another example is given by a periodic orbit *$\{S^t(u)\}_{0 \leq t \leq T}$ with period T defined by*

$$S^{t+T}(u) = S^t(u).$$

In the discrete case an element $\bar{u}$ of a periodic orbit is called periodic point *or* period q point *where q is defined by*

$$S^q \bar{u} = \bar{u}, \quad S^r \bar{u} \neq \bar{u}, \quad 0 < r < q.$$

More complicated invariant sets can be described by ω-limit sets.

Definition 1.4. We call $v \in X$ an ω-*limit point* of $u \in X$ if

$$\exists (t_i)_i, \ t_i \to \infty : S^{t_i}(u) \xrightarrow{i \to \infty} v.$$

The ω-*limit set* of $u \in X$ is defined by the union of all ω-limit points

$$\omega(u) = \{v \in X : \exists(t_i)_i, t_i \to \infty \text{ with } S^{t_i}(u) \xrightarrow{i \to \infty} v\}.$$

We extend this definition to ω-*limit sets* of bounded sets $B \subset X$ by

$$\omega(B) = \{v \in X : \exists t_i \to \infty, u_i \in B : S^{t_i}(u_i) \xrightarrow{i \to \infty} v\}.$$

Remark. *It can easily be seen that ω-limit sets are characterized by*

$$\omega(B) = \bigcap_{\substack{s \geq 0 \\ s \in \mathbb{T}}} \overline{\bigcup_{t \geq s} S^t(B)}.$$

By definition ω-limit sets are positive invariant. If the set $\bigcup_{t \geq 0} S^t(B)$ is bounded, then the ω-limit set $\omega(B)$ is also invariant.

Attractors are special invariant sets with an attracting property. These sets play an important role in the long-time behavior of dynamical systems. Some special attractors can be found via the AIM algorithm as we will see later.

Definition 1.5. A set $A \subset \mathbb{R}^N$ *attracts* $B \subset \mathbb{R}^N$ under $S : \mathbb{T} \times \mathbb{R}^N \to \mathbb{R}^N$ if for every $\varepsilon > 0$ there exists a $t^* \in \mathbb{T}$ with

$$S^t B \subset A^\varepsilon \quad \text{for all } t \geq t^*. \tag{1.3}$$

Here, A^ε denotes the ε-neighborhood of A with respect to the Euclidean distance:

$$A^\varepsilon := \{y \in \mathbb{R}^N : \|x - y\|_2 < \varepsilon \text{ for some } x \in A\}.$$

A set $A \subset \mathbb{R}^N$ is called an *attractor* if it holds

A is invariant and compact and A attracts an open neighborhood.

A is called a *global attractor* if A attracts every bounded set.

Remark. *If $S : \mathbb{T} \times \mathbb{R}^N \to \mathbb{R}^N$ is continuous and $B \subset \mathbb{R}^N$ is a bounded open set with*

$$S^t(\overline{B}) \subset B \quad \text{for all } t > 0$$

then it follows that the ω-limit set $\omega(B)$ is an attractor with representation

$$\omega(B) = \bigcap_{t \geq 0} S^t(B).$$

Hyperbolic sets

We have a closer look at fixed points of dynamical systems and define hyperbolic fixed points. Since linearizations are essential for the theory of hyperbolic sets, we have to give some definitions each for the discrete and the time continuous case.

Definition 1.6.   • A fixed point $\bar{u}$ of (1.1) is said to be *hyperbolic* if none of the eigenvalues of $DF(\bar{u})$ lies on the boundary of the unit circle. Define the subspaces $E^s, E^u \subset \mathbb{R}^N$ as the range of the projections P_s, P_u according to the spectrum Λ_s inside and Λ_u outside the unit circle:

$$\Lambda_s = \{\lambda \in \sigma(DF(\bar{u})) : |\lambda| < 1\}, \quad \Lambda_u = \{\lambda \in \sigma(DF(\bar{u})) : |\lambda| > 1\}.$$

• An equilibrium point $\bar{u}$ of (1.2) is said to be *hyperbolic* if none of the eigenvalues of $Df(\bar{u})$ lies on the imaginary axis. Define the subspaces $E^s, E^u \subset \mathbb{R}^N$ as the range of the projections P_s, P_u according to the spectrum in the left and in the right half plane

$$\Lambda_s = \{\lambda \in \sigma(DF(\bar{u})) : \Re\lambda < 0\}, \quad \Lambda_u = \{\lambda \in \sigma(DF(\bar{u})) : \Re\lambda > 0\}.$$

It follows that the state space decomposes into a direct sum of E^s and E^u for a hyperbolic fixed point:

$$\mathbb{R}^N = E^s \oplus E^u.$$

The Hartman-Grobman Theorem roughly says that the dynamics of the linearized and the nonlinear system near a hyperbolic fixed point are qualitatively the same.

Theorem 1.7 (Hartman-Grobman, discrete). *Let $F \in C^1(X, \mathbb{R}^N)$ and $\bar{u}$ be a hyperbolic fixed point of (1.1). Then there exists a $\delta > 0$, a neighborhood $\mathcal{N}$ of the origin and a homeomorphism $\mathcal{F} : B(\bar{u}, \delta) \to \mathcal{N}$ with*

$$\mathcal{F}(F(u)) = DF(\bar{u})\mathcal{F}(u) \quad \text{for all } u \in B(\bar{u}, \delta) \cap F^{-1}(B(\bar{u}, \delta)).$$

Theorem 1.8 (Hartman-Grobman, continuous). *Let $f \in C^1(X, \mathbb{R}^N)$ and $\bar{u}$ be a hyperbolic equilibrium point of (1.2). Then there exists a $\delta > 0$, a neighborhood $\mathcal{N}$ of the origin and a homeomorphism $\mathcal{F} : B(\bar{u}, \delta) \to \mathcal{N}$ such that $w(t) = \mathcal{F}(u(t))$ solves*

$$w_t = Df(\bar{u})w, \quad w(0) = w_0$$

if and only if $u(t)$ solves

$$u_t = f(\bar{u} + u), \quad u(0) = \bar{u} + w_0.$$

The dynamics around hyperbolic fixed points are described by the stable and unstable manifold:

Definition 1.9. The *unstable manifold* of a fixed point $\bar{u}$ of (1.1) and an equilibrium point of (1.2) respectively is defined as the set

$$W^u(\bar{u}) := \{u \in X : S^t(u) \xrightarrow{t \to -\infty} \bar{u}\}.$$

The *stable manifold* of $\bar{u}$ is defined as the set

$$W^s(\bar{u}) = \{u \in X : S^t(u) \xrightarrow{t \to \infty} \bar{u}\}.$$

For a hyperbolic fixed point and equilibrium point, respectively, the sets of definition 1.9 are indeed manifolds represented by the union of flows of the local stable and unstable manifold, see [SH96] for details. In this sense, the following properties of these manifolds follow from the Hartman-Grobman Theorem.

Theorem 1.10. *Let F and f be C^k-functions and $\bar{u}$ be a hyperbolic fixed point of (1.1) and an hyperbolic equilibrium point of (1.2), respectively. Then the manifolds $W^s(\bar{u})$ and $W^u(\bar{u})$ are tangent to the linear subspaces E^s and E^u at the equilibrium points and are locally representable as C^k-graphs.*

A generalization of hyperbolic fixed points is given by the definition of hyperbolic sets. In the following we will restrict ourselves to the case of a discrete dynamical system since the next definitions aim at the classification of discrete dynamical systems with nice ergodic properties. Later on we will see that on these discrete dynamical systems the convergence theory of the AIM algorithm gives satisfying results concerning the approximated invariant measures.

Definition 1.11. Let $F : X \to X$ be a C^1-diffeomorphism. A compact invariant subset $M \subset X$ is called *hyperbolic* if the tangent space $T_M(X)$ can be decomposed along M

$$T_M(X) = E_M^s \oplus E_M^u$$

and there are constants $C > 0$, $\lambda \in (0, 1)$ with the following properties for all $v \in M$:

1. $DF(v)(E_v^s) = E_{F(v)}^s$ and $DF(v)^{-1}(E_v^u) = E_{F^{-1}(v)}^u$.

2. $\|DF(v)^{-n}(w)\| \leq C\lambda^n \|w\|$ for every $w \in E_v^u$ and $n \geq 0$.

3. $\|DF(v)^n(w)\| \leq C\lambda^n \|w\|$ for every $w \in E_v^s$ and $n \geq 0$.

Next we need some topological concepts about discrete dynamical systems.

Definition 1.12. A point $v \in X$ is called *nonwandering* if for every neighborhood U of v there is a time $j \in \mathbb{N}$, $j \neq 0$, with

$$F^j(U) \cap U \neq \emptyset.$$

We denote the set of nonwandering points by $\Omega = \Omega(F)$:

$$\Omega(F) := \{v \in X : v \text{ is nonwandering}\}.$$

The dynamical system defined by F is called

- *topologically transitive* if for every $U, V \subset X$ there is a time $j > 0$ with $F^j(U) \cap V \neq \emptyset$.

- *topologically mixing* if for every $U, V \subset X$ there is a time $J > 0$ with $F^{-j}(U) \cap V \neq \emptyset$ for all $j \geq J$.

It is easy to see that $\Omega(F)$ is a closed and invariant subset of X. Hence the following definition makes sense:

Definition 1.13. A diffeomorphism $F : X \to X$ is called an *Axiom-A diffeomorphism* (or sometimes Smale diffeomorphism) if the set of nonwandering points $\Omega(F)$ is hyperbolic and the periodic points of F are dense in $\Omega(F)$:

$$\Omega(F) = \overline{\{v \in X : F^j(v) = v \text{ for a time } j > 0\}}.$$

Though the Axiom-A condition is hard to check for a given dynamical system, the theory of Axiom-A diffeomorphisms is widely spread since it offers nice spectral and—as we will see later on—ergodic properties of the underlying dynamical system. We state a result by Smale concerning the spectral decomposition of an Axiom-A diffeomorphism:

Theorem 1.14. *Let $F : X \to X$ be an Axiom-A diffeomorphism. Then the set of non-wandering points $\Omega = \Omega(f)$ can be decomposed into pairwise disjoint sets*

$$\Omega = \Omega_0 \cup \ldots \cup \Omega_{b-1},$$

with the following properties

1. *Ω_i is closed and invariant, $F_{|\Omega_i}$ is topologically transitive.*

2. *Ω_i can be decomposed into pairwise disjoint sets*

$$\Omega_i = \Omega_{i,0} \cup \ldots \cup \Omega_{i,c_i-1}$$

with the following properties

- *The sets $\Omega_{i,\ell}$ are closed and are mapped cyclically by F*

$$F(\Omega_{i,\ell}) = \Omega_{i,(l+1) \bmod c_i}. \tag{1.4}$$

- *$F^{c_i}{}_{|\Omega_{i,\ell}}$ is topologically mixing.*

Proof. [Sma67] $\square$

Definition 1.15. Let $\Omega = \Omega_0 \cup \ldots \cup \Omega_{b-1}$ be the decomposition of the set of nonwandering points of an Axiom-A diffeomorphism. We call Ω_i a *hyperbolic attractor* if it has the attracting property (1.3) for some neighborhood B.

Later on we will state an ergodic result concerning the existence of a proper invariant measure for Axiom-A diffeomorphisms. Before that we have to introduce basic definitions and properties giving statistical information about discrete dynamical systems. These properties are also called ergodic properties.

1.2 Review of basic concepts from ergodic theory

In this short introduction to ergodic theory we follow the framework of Chapter 4 in [KH95]. Proofs of the following can be found in [KH95], [Sin94] or in other standard literature about ergodic theory.

In the following we restrict ourselves to the case of discrete dynamical systems given by (1.1) where $F \in C(X, \mathbb{R}^N)$ with $F(X) \subset X$. We assume X to be a compact subset of $\mathbb{R}^N$. In particular, X is a metric space with the usual Euclidean metric.

We aim to derive statistical information about the dynamical system. Therefore a first useful term is given by the asymptotic frequency $f(u, A)$ of the orbit starting in $u \in X$.

Definition 1.16. The *asymptotic density of the distribution of the iterates of* u between $A \subset X$ and A^c is defined by the limit

$$f(u, A) := \lim_{n \to \infty} \frac{1}{n} \#\{k < n : F^k(u) \in A\}, \tag{1.5}$$

if it exists.

Using the characteristic function $\mathbb{1}_A$ of $A \subset X$ as defined in (A.8), the equation (1.5) can be written as

$$f(u, A) = \lim_{n \to \infty} \frac{1}{n} \sum_{k=0}^{n-1} \mathbb{1}_A(F^k(u)). \tag{1.6}$$

We call (1.6) the *time average* or *Birkhoff average* of $\mathbb{1}_A$ at u.

To derive results about the existence of such time averages it is useful to focus on continuous functions instead of characteristic functions. Therefore we generalize (1.6) to the following definition:

Definition 1.17. For given $u \in X$ the *time average* of $\varphi \in C(X) = C(X, \mathbb{R})$ at u is defined by

$$I_u(\varphi) = \lim_{n \to \infty} \frac{1}{n} \sum_{k=0}^{n-1} \varphi(F^k(u)) \tag{1.7}$$

if it exists.

In the proof of the next lemma we will see that the existence of I_u as a function on $C(X)$ for a given $u \in X$ implies the existence of the time averages $f(u, A)$ of characteristic functions.

The lemma gives a first result about the existence of invariant measures defined as follows:

Definition 1.18. Given $F \in C(X, \mathbb{R}^N)$ with $F(X) \subset X$, a Borel measure $\mu = \mu_{\text{inv}}$ is called *F-invariant* if for every measurable set $A \subset X$ it holds

$$\mu(F^{-1}(A)) = \mu(A).$$

Lemma 1.19. *If the asymptotic distribution* $I_u : C(X) \to \mathbb{R}$ *exists, it provides an F-invariant measure* μ_u *with the property*

$$I_u(\varphi) = \int_X \varphi \, d\mu_u \quad \text{for all } \varphi \in C(X).$$

From this lemma the following fundamental theorem follows. This theorem also holds in more general cases, e.g. on metrizable spaces. We restrict ourselves to the case of a metric space X to stay consistent with the rest of the theory.

Theorem 1.20 (Krylov-Bogolyubov). *Any continuous map $F : X \to X$ on a compact metric space has an invariant Borel probability measure.*

By the Krylov-Bogolyubov Theorem, every $F \in C(X, X)$ can be viewed as a measure-preserving transformation in the measure space given by the Borel measure of Theorem 1.20.

Definition 1.21. A transformation $F : (X, \mu) \to (X, \mu)$ of a measure space into itself is called *measure-preserving* if for every measurable set A the preimage $F^{-1}(A)$ is also measurable with
$$\mu(F^{-1}(A)) = \mu(A).$$

Theorem 1.22 (Birkhoff Ergodic Theorem). *Let X be a compact metric space, μ a Borel probability measure and $F : (X, \mu) \to (X, \mu)$ a measure-preserving transformation. If $\varphi \in L^1(X, \mu)$, then the time average*
$$\lim_{n \to \infty} \frac{1}{n} \sum_{k=0}^{n-1} \varphi(F^k(u)) =: \varphi_F(u)$$
exists for μ-almost every $u \in X$ and φ_F is a measurable function.

Remark. *The function φ_F is constructed via the Radon-Nikodym derivative of φ from which the measurability follows. Furthermore, φ_F is F-invariant since the time average is F-invariant:*
$$\varphi_F(F(u)) = I_{F(u)}(\varphi) = I_u(\varphi \circ F) = I_u(\varphi) = \varphi_F(u). \tag{1.8}$$
By the F-invariance of μ, it holds for every $n \in \mathbb{N}$
$$\int_X \frac{1}{n} \sum_{k=1}^{n} \varphi(F^k(u)) d\mu = \int_X \varphi \, d\mu$$
and hence, it follows for the limit
$$\int_X \varphi_F \, d\mu = \int_X \varphi \, d\mu. \tag{1.9}$$

Corollary 1.23. *Let X be a compact metric space and $F \in C(X, X)$. Then the set*
$$\{u \in X : \lim_{n \to \infty} \frac{1}{n} \sum_{k=1}^{n} \varphi(F^k(u)) \text{ exists for all } \varphi \in C(X)\}$$
has full measure with respect to any F-invariant Borel probability measure.

Definition 1.24. An F-invariant measure μ is called *ergodic* with respect to F if for every F-invariant measurable set $A \subset X$ it holds
$$\mu(A) \in \{0, 1\}.$$
Alternatively, we say F is ergodic with respect to μ.

The ergodicity of a measure space can be reformulated in terms of a functional property

Proposition 1.25. *If $F : (X, \mu) \to (X, \mu)$ is ergodic, every measurable F-invariant function is constant μ-almost everywhere.*

The proposition implies an important corollary of the Birkhoff Ergodic Theorem:

Corollary 1.26. *If $F : X \to X$ is an ergodic μ-preserving transformation where μ is a probability measure and $\varphi \in L^1(X, \mu)$ then for μ-almost every $u \in X$ it follows*

$$\varphi_F(u) = \lim_{n \to \infty} \frac{1}{n} \sum_{k=0}^{n-1} \varphi(F^k(u)) = \int_X \varphi \, d\mu \tag{1.10}$$

The Birkhoff property (1.10) shows the importance of invariant measures when exploring the long-time behavior of dynamical systems. It states that for μ-almost every $u \in X$ the time average (1.7) of a trajectory starting in u is given by the spatial distribution defined by the ergodic measure μ, also called the *space average* of the dynamical system.

However, without further assumptions it is likely that the ergodic measure μ is degenerated in the sense that it is not continuous with respect to the Lebesgue measure λ on X. In that case the time-space equality (1.10) only holds on a Lebesgue null set, i.e. on a numerical irrelevant set.

If we postulate higher regularity on our dynamical system we will see that we can strengthen the equality (1.10) to a set of initial points with positive Lebesgue measure. A measure μ with this property is called SRB measure named after Sinai, Ruelle and Bowen.

Definition 1.27. An invariant measure μ is called an *SRB-measure* if there exists a measurable set $A \subset X$ with positive Lebesgue measure such that time average equals space average on A:

$$\lim_{n \to \infty} \frac{1}{n} \sum_{k=0}^{n-1} \varphi(F^k(u)) = \int_X \varphi \, d\mu \quad \text{for all } \varphi \in C(X), u \in A. \tag{1.11}$$

The ergodicity of an SRB measure follows immediately. In Definition 1.13 we have defined dynamical systems for which SRB measures exist:

Theorem 1.28 (Ruelle). *Let $F : X \to X$ be an Axiom-A diffeomorphism. Let $\Omega \subset \Omega(F)$ be a hyperbolic attractor in terms of definition 1.15. Then there exists an SRB measure μ_Ω with the following properties: There is a neighborhood U of Ω with positive Lebesgue measure such that time average equals space average on U, i.e. for every $u \in U$ it holds:*

$$\lim_{n \to \infty} \frac{1}{n} \sum_{k=0}^{n-1} \varphi(F^k(u)) = \int_X \varphi \, d\mu_\Omega. \quad \text{for all } \varphi \in C(X) \tag{1.12}$$

Proof. [Rue76]   □

1.3   Ulam's method

The following approach for the approximation of invariant measures in discrete dynamical systems goes back to Ulam [Ula60].

1. Define an operator P on the space of probability measures $\mathcal{M}$ on X such that fixed points of P are precisely given by invariant measures.

2. Discretize P by a Galerkin approach with an adequate system of finite subspaces of $\mathcal{M}$ and get a nonnegative matrix P_k as a discretization of P.

3. Approximate invariant measures by fixed points of the nonnegative matrix P_k where the existence is given by the Perron-Frobenius theory.

In fact, Ulam formulated this approach for a dynamical system defined on the unit interval. For a given partition of $(0, 1)$ into intervals, discrete densities are given by according step functions. He supposed that under proper assumptions the fixed points of the matrix P_k should converge to an invariant measure in the L^1 sense.

Since then many authors have worked on generalizations and convergence properties of this algorithm, among those Li ([Li76]), Keller ([Kel82]), Ding et al. ([DDL93], [DZ96]) and Froyland ([Fro95]). In the following we work out the ansatz of Dellnitz, Junge et al. which is the basis for the subdivision algorithm and described in e.g. [DJ99], [DHJR97] and [DFJ01].

We start with some definitions and notions needed to apply Ulam's method in our context. As before, $X \subset \mathbb{R}^N$ shall be compact and $F \in C(X, X)$ defines a dynamical system via (1.1). Let $\mathcal{M}^1(X)$ be the set of probability measures μ on X. Recall that $\mathcal{M}^1(\mathbb{R}^N)$ is a compact metric space with the weak metric, see A.38.

Definition 1.29. Let $F : X \to X$ be a diffeomorphism. The *Frobenius-Perron operator* on $\mathcal{M}^1(\mathbb{R}^N)$ is defined by

$$P\mu(A) = \mu(F^{-1}(A)), \quad \text{for all measurable } A \subset X. \tag{1.13}$$

On the Lebesgue space $L^1(X) = L^1(X, \lambda_N)$ according to the N-dimensional Lebesgue measure λ_N, the Frobenius-Perron operator $P : L^1(X) \to L^1(X)$ is defined by

$$\int Pg \, d\lambda_N = \int_{F^{-1}(A)} g \, d\lambda_N \quad \text{for all measurable } A \subset X.$$

For the discretization of P consider a finite *partition* $\mathcal{B}_k$ of X, i.e.

$$\mathcal{B}_k = \{B_1, \dots, B_K\}$$

where the partition sets $B_i \subset X$ have positive Lebesgue measure, the sets are disjoint in the Lebesgue sense ($\lambda_N(B_i \cap B_j) = 0$ for $i \neq j$) and they cover X ($\lambda_N(\bigcup B_i) = \lambda_N(X)$). On such a partition we define a finite-dimensional subspace $V_k \subset L^1(X, \lambda_N)$ by

$$V_k := \{g : X \to \mathbb{R} : h_{|B_i} \text{ is constant for every } i = 1, \dots, K\}$$
$$= \text{span} \left\{ \frac{\mathbb{1}_{B_1}}{\lambda_N(B_1)}, \dots, \frac{\mathbb{1}_{B_K}}{\lambda_N(B_k)} \right\}. \tag{1.14}$$

It is easy to see that the orthogonal projection $Q_k : L^1(X, \lambda_N) \to V_k$ onto such a subspace, called the Galerkin projection, is given by

$$Q_k(g) = \sum_{i=1}^{K} \left(\frac{1}{\lambda_N(B_i)} \int_{B_i} g \, d\lambda_N \right) \mathbb{1}_{B_i}.$$

It follows that the discretized Frobenius-Perron operator $P_k = Q_k \circ P_{|V_k} : V_k \to V_k$ can be identified with the matrix $P_k = (p_{ij})_{ij} \in \mathbb{R}^{K,K}$ with entries

$$p_{ij} = \frac{\lambda_N \left(B_j \cap F^{-1}(B_i) \right)}{\lambda_N(B_j)}.$$

Since F maps X into itself, the nonnegative matrix $P_k \in \mathbb{R}^{K,K}$ is *stochastic* in the sense that all column sums of P_k equal 1. By the Perron-Frobenius Theorem A.2 the existence of a nonnegative fixed point $v_k \in \mathbb{R}^K$, $P_k v_k = v_k$ follows. This fixed point defines a density

$$h_k = \sum_{i=1}^{K} (v_k)_i \frac{\mathbb{1}_{B_i}}{\lambda_N(B_i)} \in V_k$$

corresponding to the so-called discrete measure μ_k with $\mu_k(A) = \int_A h_k \, d\lambda_N$. A representation of μ_k in terms of the Lebesgue measure can easily be derived:

$$\mu_k(A) = \sum_{i=1}^{K} (v_k)_i \frac{\lambda_N(A \cap B_i)}{\lambda_N(B_i)} \quad \text{for all measurable } A \subset X. \tag{1.15}$$

1.4 The adaptive invariant measure (AIM) algorithm

Using the approximation theory from the last section we introduce the following subdivision algorithm developed by Dellnitz, Junge and co-authors. The algorithm computes a sequence of pairs $(\mathcal{B}_k, \mu_k)$ consisting of partitions $\mathcal{B}_k$ of (a subset of) X and corresponding discrete measures μ_k. For simplicity we assume that there exists a positive invariant generalized rectangle

$$B_0^{(N)} = B(c_0, r_0) = \{x \in \mathbb{R}^N : |x_i - (c_0)_i| \leq (r_0)_i, i = 1, \dots, N\} \subset X \tag{1.16}$$

from which all partition sets arise by bisection. Therefore all partition sets are also generalized rectangles and we call them simply boxes in the following.

An essential feature of the algorithm is a flexible bisection process adapted to the given dynamical system by use of the discrete measures μ_k. Without this adaptation, the regions of substantial dynamics would not be found in most applications.

Algorithm 1.30. The Adaptive Invariant Measure (AIM) algorithm

- **Initialization:** Let $B_0^{(N)}$ be a positive invariant box with center $c_0 \in \mathbb{R}^N$ and radius $r_0 \in \mathbb{R}^N$ given by (1.16). Starting with the initial box collection $\mathcal{B}_0 = \{B_0\}$, define the initial discrete measure $\mu_0 : \mathcal{B}(\mathbb{R}^N) \to [0,1]$ by

$$\mu_0(A) = \frac{\lambda_N(A \cap B_0)}{\lambda_N(B_0)}, \quad A \in \mathcal{B}(\mathbb{R}^N).$$

- **Recursion step:** Assume that a partition $\mathcal{B}_{k-1}$ of a subset of $B_0^{(N)}$ is given with a discrete measure $\mu_{k-1} : \mathcal{B}(\mathbb{R}^N) \to [0,1]$.

1. *Subdivision:* For a given number $\delta_k > 0$ choose a subset $\mathcal{B}^{(1)}$ of $\mathcal{B}_{k-1}$ by

$$\mathcal{B}^{(1)} = \{B \in \mathcal{B}_{k-1} : \mu_{k-1}(B) \geq \delta_k\}. \tag{1.17}$$

Subdivide all boxes in $\mathcal{B}^{(1)}$ in coordinate $k \mod N$ into a refined box collection $\mathcal{B}^{(2)}$ and continue with

$$\widehat{\mathcal{B}}_k := (\mathcal{B}_{k-1} \setminus \mathcal{B}^{(1)}) \cup \mathcal{B}^{(2)}, K := |\widehat{\mathcal{B}}_k|.$$

2. *Perron root:* Calculate a normalized fixed point $u \in \mathbb{R}^K$, $\|u\|_1 = 1$ of the Perron-Frobenius matrix $P_k = (p_{ij})_{ij} \in \mathbb{R}^{K,K}$ defined by

$$p_{ij} = \frac{\lambda_N(B_j \cap F^{-1}(B_i))}{\lambda_N(B_j)}, \quad 1 \leq i,j \leq K \tag{1.18}$$

where $\widehat{\mathcal{B}}_k = \{B_1, \ldots, B_K\}$.

3. *Discrete measure:* Set $\mathcal{B}_k = \{B_i \in \widehat{\mathcal{B}}_k : i = 1, \ldots, K$ and $u_i > 0\} \subset \widehat{\mathcal{B}}_k$.
   A new discrete measure $\mu_k : \mathcal{B}(\mathbb{R}^N) \to [0,1]$ is defined by

$$\mu_k(A) = \sum_{i=1}^{K} u_i \frac{\lambda_N(A \cap B_i)}{\lambda_N(B_i)}, \quad A \in \mathcal{B}(\mathbb{R}^N), \quad B_i \in \mathcal{B}_k.$$

Remark. *By definition the discrete measures are probability measures and in particular* $\mu_k(X) = 1$.
In equation (1.17) we have some freedom in choosing the adaptation constant $\delta_k > 0$. It is obvious that the adaption constants should decrease with increasing number of boxes. A common choice is the test with the average measure, i.e.

$$\delta_k = (\#\mathcal{B}_k)^{-1} \sum_{i=1}^{K} \mu_k(B_i) = (\#\mathcal{B}_k)^{-1}.$$

Since the support sets B_k are given by generalized rectangles built by bisections of the starting box $B_0^{(N)}$, they can be efficiently coded in a binary tree where every subdivision step is given by increasing the tree depth by one.
For the computation of the Lebesgue measure in (1.18), a Monte-Carlo approach *is used with proper test points $t_{j,m}$, $m = 1, \ldots, M$ in the box $B_j \in \mathcal{B}_k$. Then the matrix entry $p_{ij} \geq 0$ is approximated by*

$$
\begin{aligned}
p_{ij} &= \frac{\lambda_N(B_j \cap F^{-1}(B_i))}{\lambda_N(B_j)} = \frac{1}{\lambda_N(B_j)} \int_{B_j} \mathbb{1}_{F^{-1}(B_i)} \, d\lambda_N \\
&\approx \frac{1}{M} \sum_{m=1}^{M} \mathbb{1}_{F^{-1}(B_i)}(t_{j,m}) = \frac{1}{M} \sum_{m=1}^{M} \mathbb{1}_{B_i}(F(t_{j,m})) \\
&= \frac{1}{M} \#\{F(t_{j,m}) \in B_i : m = 1, \ldots, M\}.
\end{aligned}
\tag{1.19}
$$

Details of the storing algorithm, the Monte-Carlo approach and other implementational details can be found in [DHJR97] and [DJ02].

1.5 Convergence of the AIM algorithm

A natural question already arising within these first ideas by Ulam is whether the discretized measures μ_k defined by (1.15) converge to some invariant measure μ of (1.1):

$$\mu_k \xrightarrow{k \to \infty} \mu.$$

We focus on two considerations:

1. What kind of topology do we choose for the limit process?

2. What properties does the limit measure μ possess?

In section 1.2 we have worked out that an invariant measure gives numerically relevant statistical information about the dynamical system only if it has the SRB property (1.11).

Dellnitz, Junge et al. ([DJ02], [DFJ01]) construct a limit process such that the limit measure is an SRB measure μ_Ω as defined in (1.12). This measure, however, is no more the limit of the fixed points of P_k, the deterministic Galerkin projections of the Frobenius-Perron operator $P : \mathcal{M} \to \mathcal{M}$. Instead, the concept of stochastic processes, in particular small random perturbations, is used with resulting perturbed invariant measures μ_ε, $\varepsilon > 0$. Then, for discretizations $\mu_k(\varepsilon)$ of μ_ε, Dellnitz and Junge show convergence to the SRB measure associated to a hyperbolic attractor as in (1.12). The convergence result is stated in detail in Corollary 1.33. We will give a brief summary of the ideas of the proof.

Definition 1.31. A function $p : X \times \mathcal{B}(X) \to [0,1]$ where $\mathcal{B}(X)$ denotes the Borel-σ-algebra of X is called a *stochastic transition function* if

- $p(u, \cdot)$ is a probability measure for all $u \in X$,

- $p(\cdot, A)$ is a measurable function for all $A \in B(X)$.

The *i-step transition function* $p^i : X \times \mathcal{B}(X) \to [0,1]$ is defined by

$$p^{i+1}(u, A) = \int p^i(v, A) \, p(u, dv).$$

A probability measure $\mu \in \mathcal{M}$ is called p-invariant if $\mu(A) = \int p(u, A) \, d\mu$. The Frobenius-Perron operator related to p is defined by

$$P_p \mu(A) = \int p(u, A) \, d\mu.$$

Remark. *It is easy to show that when p is a stochastic transition function, so is p^i, $i \in \mathbb{N}$. Using the Frobenius-Perron operator we can formulate the Markov process by*

$$\mu_{i+1} = P_p \mu_i, \quad i \in \mathbb{N}. \tag{1.20}$$

More details about stochastic processes and in particular transition functions can be found in e.g. [Doo67]. Here we just note that by choosing $p(u, A) = \delta_{F(u)}(A)$ the deterministic concept of section 1.2 is embedded into this stochastic concept.

If p is absolutely continuous, i.e. $p(u, \cdot)$ is absolutely continuous with respect to the Lebesgue measure for all $u \in X$, we can write $p(u, \cdot)$ as $p(u, A) = \int_A q(u, v) \, dv$ and get an L^1-version of the Frobenius-Perron operator $P_p : L^1(X) \to L^1(X)$ by

$$P_p h = \int q(u, v) h(v) \, dv.$$

As in Section 1.3 we define the discretized operator $P_k = P_k(p)$ to a partition $\mathcal{P}_k$ of X using the Galerkin projection Q_k onto the finite space V_k, cf. (1.14), and get

$$P_k = Q_k \circ P_{p|V_k}. \tag{1.21}$$

If the operator $P_p : L^1(x) \to L^1(X)$ is compact it is easy to see that for the discrete operators $P_k = P_k(p)$ on partitions $\mathcal{P}_k$ with diam $\mathcal{P}_k \xrightarrow{k \to \infty} 0$, we get convergence in the operator norm

$$\|P_p - P_k(p)\|_{L^1} \xrightarrow{k \to \infty} 0. \tag{1.22}$$

According to the approximation theory for compact operators by Osborn ([Osb75]), the compactness of the Frobenius-Perron operator P_p together with (1.22) implies a convergence result for the fixed points h_k of P_k and h_p of P_p:

$$\|h_k - h_p\|_{L^1} \xrightarrow{k \to \infty} 0.$$

Now Dellnitz and Junge use the properties of a special family $\{p_\varepsilon\}_\varepsilon$ of transition functions, such that certain regularity assumptions hold. These are called assumption K in the convergence theory by Kifer. For details see [Kif86]. In particular, these assumptions imply the following properties:

- The Markov process (1.20) defined by $\{p_\varepsilon\}_\varepsilon$ is a *small random perturbation*, i.e.

$$p_\varepsilon(u, \cdot) \xrightarrow{\varepsilon \to 0} \delta_{F(u)} \quad \text{for } \varepsilon \to 0.$$

- p_ε is absolutely continuous and the corresponding Frobenius-Perron operator $P_\varepsilon = P_{p_\varepsilon}$ is compact, hence Osborn's approximation results hold.

- The following convergence result between the stochastic and deterministic setting holds:

Theorem 1.32 (Kifer). *Let $F : X \to X$ be an Axiom-A diffeomorphism with a hyperbolic attractor Ω and let an open set $U \supset \Omega$ be given such that the densities q_ε of p_ε satisfy*

$$q_\varepsilon(u, v) = 0 \quad \text{for all } u \in \overline{F(U)}, v \notin U.$$

Then the transition function p_ε has a unique invariant measure μ_ε with support in U and the measures μ_ε converge to the SRB measure μ_Ω of Theorem 1.28 in the weak metric:

$$\lim_{\varepsilon \to 0} \mu_\varepsilon = \mu_\Omega.$$

With the compactness of P_ε the main theorem follows as a corollary:

Corollary 1.33. *Let small random perturbations $\{p_\varepsilon\}_\varepsilon$ be given satisfying the assumptions of Theorem 1.32. Let the discrete operators $P_{k,\varepsilon} = P_k(p_\varepsilon) : L^1 \to L^1$ be defined by (1.21) on partitions $\mathcal{P}_k$ of U. Let h_k^ε be a fixed point of $P_{k,\varepsilon}$. Then the discrete measures*

$$\mu_k(\varepsilon) = \int h_k^\varepsilon d\lambda_N$$

converge in the weak metric to the SRB measure μ_Ω corresponding to the hyperbolic attractor Ω in the following sense:

$$\lim_{\varepsilon \to 0} \lim_{k \to \infty} \mu_k(\varepsilon) = \mu_\Omega. \tag{1.23}$$

Chapter 2

Model reduction and long-time behavior

In this chapter, we focus on model reduction methods. We introduce the proper orthogonal decomposition (POD) method and state the standard error analysis of this method. The relation to singular value decomposition is essential for this error analysis. We state results of Kunisch and Volkwein, [KV01], concerning the POD approximations of finite-time trajectories in parabolic systems.

Then we consider the long-time behavior of the POD method. We derive explicit error bounds in some characteristic model cases of dynamical systems. Our results are illustrated by a number of numerical experiments.

2.1 Proper orthogonal decomposition (POD)

The concept of proper orthogonal decomposition is used to produce reduced-order models mainly in problems arising in control theory. The idea is to determine a nested family of subspaces in the original state space that optimally span the data consisting of given snapshots. Usually these snapshots are derived from trajectories of the system.

We start with the definition of a POD basis in an abstract setting. Therefore, let H be a separable real Hilbert space with inner product $\langle \cdot, \cdot \rangle$.

Definition 2.1. Let a collection of snapshots in H be given by

$$\{y_i : i = 1, \ldots, m\}.$$

An ℓ-dimensional orthonormal system $\{w_k\}_{k=1}^{\ell}$ is called *proper orthogonal decomposition basis of rank ℓ corresponding to* $\{y_i\}_i$ if it solves the minimization problem

$$E_{\text{pod}}(\{\psi_k\}_{k=1}^{\ell}) := \frac{1}{m} \sum_{j=1}^{m} \|y_j - P_\psi y_j\|^2 \overset{!}{=} \min \tag{2.1}$$

where $P_\psi : H \to H$ is the orthogonal projection onto $S_\psi = \text{span}\{\psi_1, \ldots, \psi_\ell\}$:

$$P_\psi y = \sum_{k=1}^{\ell} \langle y, \psi_k \rangle \psi_k.$$

2.1.1 Singular value decomposition and the relation to POD

In order to derive an abstract result about the POD basis of rank ℓ, we introduce the singular value decomposition for compact operators in Hilbert spaces.

Theorem 2.2. *Let G, H be arbitrary Hilbert spaces and $Y : H \to G$ a compact operator with adjoint operator Y^*. Let the eigenvalues of the self-adjoint positive semidefinite operator $K = Y^*Y : H \to H$ be denoted by $\{\lambda_i\}_{i \in I}$ with*

$$\lambda_1 \geq \lambda_2 \geq \ldots \geq 0$$

where I may be finite ($I \subset \mathbb{N}$) or infinite ($I = \mathbb{N}$). Then there are orthonormal systems $\{u_i\}_{i \in I} \subset H$, $\{v_i\}_{i \in I} \subset G$ with

$$Yu_i = \sigma_i v_i \quad and \quad Y^*v_i = \sigma_i u_i$$

where $\sigma_i = \sqrt{\lambda_i}$, $i \in I$. For every $u \in H$ we get the expression

$$u = u_0 + \sum_{i \in I} \langle u, u_i \rangle u_i$$

with $u_0 \in ker(K)$ and

$$Ku = \sum_{i \in I} \sigma_i \langle u, u_i \rangle v_i. \tag{2.2}$$

Proof. The proof can be found in [Kir96] or [Lou89].   □

As in the finite-dimensional case (see Appendix) we call the decomposition given by (2.2) singular value decomposition.

Definition 2.3. For a compact operator $Y : H \to G$ between Hilbert spaces as above the *singular value decomposition* of Y is defined by $\{\sigma_i, u_i, v_i\}_{i \in I}$ as given by Theorem 2.2. More precisely, we call σ_i the *singular values*, u_i the *right* and v_i the *left singular vectors*.

The POD method in a Hilbert space H can be described in terms of the singular value decomposition of a suitable linear operator. Therefore we consider the linear operator $Y : \mathbb{R}^m \to H$ defined for the given collection of snapshots $\{y_i\}_{i \leq m}$ by

$$Y(w) = \frac{1}{\sqrt{m}} \sum_{i=1}^{m} w_i y_i. \tag{2.3}$$

Observe that the range of Y is finite, hence Y is a compact operator and we can use the singular value decomposition of Y. The adjoint is given by $Y^* : H \to \mathbb{R}^m$ with

$$Y^*(\varphi) = \frac{1}{\sqrt{m}}(\langle \varphi, y_1 \rangle, \ldots, \langle \varphi, y_m \rangle)^T. \tag{2.4}$$

In our case the self-adjoint operator $K = Y^*Y : \mathbb{R}^m \to \mathbb{R}^m$ can be identified with the matrix

$$K = \frac{1}{m}((\langle y_j, y_i \rangle))_{ij} \in \mathbb{R}^{m,m}.$$

K is called *correlation matrix*. Now we denote the right singular vectors as defined in Theorem 2.2 by $v_k \in \mathbb{R}^m$ and the left singular vectors by $w_k \in H$. Observe that the right singular vectors are the eigenvectors of the correlation matrix. Using the fact

$$Y v_k = \sigma_k w_k,$$

the left singular vectors can be expressed by

$$w_k = \frac{1}{\sqrt{m}\sigma_k} \sum_{i=1}^{m} (v_k)_i y_i. \tag{2.5}$$

Theorem 2.4. *Let a collection $\{y_i\}_{i=1}^m$ of snapshots be given in a separable Hilbert space H. The corresponding linear operator $Y : \mathbb{R}^m \to H$ shall be given by (2.3). Let $d \leq m$ be the dimension of the subspace spanned by the collection of snapshots $\mathrm{span}\{y_1, \ldots, y_m\}$. Then the POD basis of rank $\ell \leq d$ is given by the left singular vectors $\{w_k\}_{k=1}^\ell$ of Y.*

The approximation error is given by the singular values of Y:

$$E_{\mathrm{pod}}(\{w_k\}_{k=1}^\ell) = \sum_{k=\ell+1}^{d} \sigma_k^2.$$

In the finite-dimensional case $H = \mathbb{R}^N$ the operator $Y : \mathbb{R}^m \to \mathbb{R}^N$ is given by a matrix and we get the following result

Corollary 2.5. *Let a collection $\{y_i\}_{i=1}^m$ of snapshots be given in $H = \mathbb{R}^N$. Then the POD basis of rank ℓ is given by the singular vectors $w_k \in \mathbb{R}^N$ of*

$$Y = \frac{1}{\sqrt{m}} col(y_1, \ldots, y_m) \in \mathbb{R}^{N,m}$$

where $col(y_1, \ldots, y_m)$ denotes the matrix with columns $y_1, \ldots, y_m$. The approximation error is given by the sum of the remaining squared singular values:

$$E_{\mathrm{pod}}(\{w_k\}_{k=1}^\ell) = \sum_{k=\ell+1}^{d} \sigma_k^2.$$

For the proof of Theorem 2.4 we need a variational lemma about self-adjoint operators in Hilbert spaces

Lemma 2.6. *Let $T : H \to H$ be a positive semidefinite, self-adjoint and compact operator in a separable Hilbert space H with eigenpairs $(\lambda_k, \varphi_k)_k$ where*

$$\{\varphi_k\}_{k=1}^\infty \text{ is an orthonormal system, } \lambda_1 \geq \lambda_2 \geq \ldots \geq 0.$$

Then for every $\ell \in \mathbb{N}$ the system $\{\varphi_k\}_{k=1}^\ell$ has the following maximizing property:

$$\max\{\sum_{k=1}^{\ell} \langle T\psi_k, \psi_k \rangle : \{\psi_k\}_{k=1}^\ell \text{ orthonormal}\} = \sum_{k=1}^{\ell} \langle T\varphi_k, \varphi_k \rangle = \sum_{k=1}^{\ell} \lambda_k.$$

Proof. By Theorem A.14 we have for every $x \in H$

$$Tx = \sum_{k=1}^{\infty} \lambda_k \langle x, \varphi_k \rangle \varphi_k$$

It follows immediately that

$$\sum_{k=1}^{\ell} \langle T\varphi_k, \varphi_k \rangle = \sum_{k=1}^{\ell} \lambda_k \langle \varphi_k, \varphi_k \rangle = \sum_{k=1}^{\ell} \lambda_k.$$

Let $\psi_1, \ldots, \psi_\ell$ be an arbitrary orthonormal system. Denote by $S_\varphi, S_\psi \subset H$ the subspaces spanned by $\varphi_1, \ldots, \varphi_\ell$ and $\psi_1, \ldots, \psi_\ell$, respectively. Then we consider two cases:

a) $S_\psi \cap S_\varphi^\perp \neq \{0\}$: We can write ψ in terms of the ONS of T:

$$\psi_i = \sum_{j=1}^{\infty} \langle \psi_i, \varphi_j \rangle \varphi_j + z_i, i = 1, \ldots, \ell$$

with $z_i \in \ker(T)$, $i = 1, \ldots, \ell$. Let P_ψ be the orthogonal projection onto S_ψ, i.e.

$$P_\psi \varphi = \sum_{k=1}^{\ell} \langle \varphi, \psi_k \rangle \psi_k, \quad \varphi \in H.$$

It follows for $q_{ij} := |\langle \varphi_j, \psi_i \rangle|$:

$$1 = \|\varphi_j\|^2 \geq \|P_\psi \varphi_j\|^2 = \sum_{i=1}^{\ell} |\langle \varphi_j, \psi_i \rangle|^2 = \sum_{i=1}^{\ell} q_{ij}^2, \quad j \in \mathbb{N} \tag{2.6}$$

and

$$1 = \|\psi_i\|^2 \geq \sum_{j=1}^{\infty} |\langle \varphi_j, \psi_i \rangle|^2 = \sum_{j=1}^{\infty} q_{ij}^2, \quad i = 1, \ldots, k.$$

$$\Rightarrow \sum_{j=\ell+1}^{\infty} q_{ij}^2 = \sum_{j=1}^{\infty} q_{ij}^2 - \sum_{j=1}^{\ell} q_{ij}^2 \leq 1 - \sum_{j=1}^{\ell} q_{ij}^2.$$

Summing up over $i = 1, \ldots, \ell$ leads to

$$\sum_{i=1}^{\ell} \sum_{j=\ell+1}^{\infty} q_{ij}^2 \leq \ell - \sum_{i=1}^{\ell} \sum_{j=1}^{\ell} q_{ij}^2 = \sum_{j=1}^{\ell} \left(1 - \sum_{i=1}^{\ell} q_{ij}^2 \right) \tag{2.7}$$

with $1 - \sum_{i=1}^{\ell} q_{ij}^2 \geq 0$ following from (2.6). Combining all inequalities, we get the desired result:

$$
\begin{aligned}
\sum_{k=1}^{\ell} \langle T\psi_k, \psi_k \rangle
&= \sum_{k=1}^{\ell} \sum_{j=1}^{\infty} \lambda_j \langle \psi_k, \varphi_j \rangle \langle \varphi_j, \psi_k \rangle \\
&= \sum_{k=1}^{\ell} \sum_{j=1}^{\infty} \lambda_j q_{kj}^2 \\
&\leq \sum_{k=1}^{\ell} \left(\sum_{j=1}^{\ell} \lambda_j q_{kj}^2 + \lambda_{\ell+1} \sum_{j=\ell+1}^{\infty} q_{kj}^2 \right) \\
&= \sum_{k=1}^{\ell} \sum_{j=1}^{\ell} \lambda_j q_{kj}^2 + \lambda_{\ell+1} \sum_{k=1}^{\ell} \sum_{j=\ell+1}^{\infty} q_{kj}^2 \\
&\overset{(2.7)}{\leq} \sum_{k=1}^{\ell} \sum_{j=1}^{\ell} \lambda_j q_{kj}^2 + \sum_{j=1}^{\ell} \underbrace{\lambda_{\ell+1}}_{\leq \lambda_j} \underbrace{(1 - \sum_{k=1}^{\ell} q_{kj}^2)}_{\geq 0} \\
&\leq \sum_{j=1}^{\ell} \left(\sum_{k=1}^{\ell} \lambda_j q_{kj}^2 + \lambda_j (1 - \sum_{k=1}^{\ell} q_{kj}^2) \right) = \sum_{j=1}^{\ell} \lambda_j.
\end{aligned}
$$

b) $S_\psi = S_\varphi$: We have

$$
\psi_i = \sum_{j=1}^{\ell} q_{ij} \varphi_j, \quad i = 1, \dots, \ell
$$

with $q_{ij} = \langle \psi_i, \varphi_j \rangle$ as above. Since $\{\psi_i\}_{i=1}^{\ell}$ is orthonormal it follows immediately that the rows of $Q = (q_{ij})_{ij} \in \mathbb{R}^{\ell,\ell}$ are orthonormal since

$$
\delta_{k\ell} = \langle \psi_k, \psi_\ell \rangle = \sum_{j=1}^{\ell} q_{kj} q_{\ell j}.
$$

That means Q is an orthogonal matrix and, in particular, also the columns of Q are orthonormal which shows that $\{\psi_i\}_{i=1}^{\ell}$ leads to the same value for the induced scalar product

$$
\sum_{k=1}^{\ell} \langle T\psi_k, \psi_k \rangle = \sum_{k=1}^{\ell} \sum_{j=1}^{\ell} \lambda_j \underbrace{q_{kj} q_{kj}}_{=1} = \sum_{j=1}^{\ell} \lambda_j.
$$

$\square$

Proof of Theorem 2.4. First observe that the minimization problem (2.1) is equivalent to maximizing

$$
\frac{1}{m} \sum_{j=1}^{\ell} \| P_\psi y_j \|^2
$$

under orthonormal bases $\{\psi_k\}_{k=1}^{\ell}$ since

$$E_{\text{pod}}(\{\psi_k\}_{k=1}^{\ell}) = \frac{1}{m}\sum_{j=1}^{m}\|y_j - P_{\psi}y_j\|^2$$

$$= \frac{1}{m}\sum_{j=1}^{m}\|y_j\|^2 - \frac{1}{m}\sum_{j=1}^{m}\|P_{\psi}y_j\|^2.$$

We define $R = YY^* : H \to H$ and get

$$R(\varphi) = \frac{1}{m}\sum_{j=1}^{m}\langle\varphi, y_j\rangle y_j.$$

By definition R is self-adjoint and compact since Y is compact. The eigenvectors of R are the left singular vectors of Y. Hence by Lemma 2.6, the left singular vectors of Y solve the minimization problem (2.1).

For the error bound we use the trace property: $\text{trace}(A) = \|A\|_F^2$ for square matrices A. Applied to $K = Y^*Y$ we get

$$\sum_{j=1}^{m}\|y_j\|^2 = \|K\|_F^2 = \text{trace}(K) = \sum_{k=1}^{d}\lambda_k.$$

It follows for the error

$$E_{\text{pod}}(\{\varphi_k\}_{k=1}^{\ell}) = \frac{1}{m}\left(\sum_{j=1}^{m}\|y_j\|^2 - \sum_{j=1}^{m}\|P_{\varphi}y_j\|^2\right) = \frac{1}{m}\sum_{k=\ell+1}^{d}\lambda_k.$$

$\square$

2.1.2   Perturbation theory for the Singular Value Decomposition

In Section 2.3 we will analyze the POD error using perturbation results for the singular value decomposition. A standard result for the approximation of perturbed singular values is based on the Theorem of Mirsky (A.4). It treats the perturbation of eigenvalues of symmetric matrices and can be used for singular values. Since we can extend this result to the case of snapshots in an arbitrary, not necessarily finite dimensional Hilbert space, we state the result here.

Theorem 2.7. *In $H = \mathbb{R}^N$ let two collections $\{y_i\}_i$ and $\{\tilde{y}_i\}_i$ of snapshots be given. As in Corollary 2.5 we arrange these snapshots in matrices*

$$Y = \frac{1}{\sqrt{m}}\,\text{col}(y_1,\ldots,y_m), \quad \tilde{Y} = \frac{1}{\sqrt{m}}\,\text{col}(\tilde{y}_1,\ldots,\tilde{y}_m).$$

Then the distance between the singular values σ_i of Y and $\tilde{\sigma}_i$ of $\tilde{Y}$ satisfies

$$\sum_{i=1}^{m}(\sigma_i - \tilde{\sigma}_i)^2 \le \frac{1}{m}\sum_{i=1}^{m}\|y_i - \tilde{y}_i\|_2^2. \tag{2.8}$$

Proof. By Theorem A.3 the eigenvalues of the symmetric matrix $\begin{pmatrix} 0 & Y^T \\ Y & 0 \end{pmatrix}$ are given by $\sigma_1 \geq \ldots \geq \sigma_m \geq 0 \geq -\sigma_m \geq \ldots \geq -\sigma_1$. The same holds for the matrix $\tilde{Y}$ and the corresponding singular values $\tilde{\sigma}_i$. By the Theorem of Mirsky it follows

$$2 \sum_{i=1}^{m} (\sigma_i - \tilde{\sigma}_i)^2 = 2\| \operatorname{diag}(\{\sigma_i\}_i) - \operatorname{diag}(\{\tilde{\sigma}_i\}_i)\|_F^2$$

$$= \| \operatorname{diag}(\{\sigma_i - \tilde{\sigma}_i\}_i, \{-(\sigma_i - \tilde{\sigma}_i)\}_i, 0)\|_F^2$$

$$\leq \| \begin{pmatrix} 0 & E^T \\ E & 0 \end{pmatrix} \|_F^2 = 2\|E\|_F^2$$

where $E = Y - \tilde{Y}$. Moreover, by definition,

$$\|E\|_F^2 = \|Y - \tilde{Y}\|_F^2 = \|\frac{1}{\sqrt{m}} \operatorname{col}(y_1 - \tilde{y}_1, \ldots, y_m - \tilde{y}_m)\|_F^2 = \frac{1}{m} \sum_{i=1}^{m} \|y_i - \tilde{y}_i\|_2^2.$$

$\square$

Now let $(H, \langle \cdot, \cdot \rangle)$ be an arbitrary separable real Hilbert space and two collections of snapshots be given by $\{y_i\}_{i \leq m}$ and $\{\tilde{y}_i\}_{i \leq m}$ in H. We define the corresponding operators $Y, \tilde{Y} : \mathbb{R}^m \to H$ as in (2.3). We have already seen that the adjoints $Y^*, \tilde{Y}^* : H \to \mathbb{R}^m$ of the compact operators $Y, \tilde{Y}$ can be defined by (2.4). As in the finite-dimensional case, we have seen that the error of the proper orthogonal decomposition is given by the singular values of $Y : \mathbb{R}^m \to H$ and $\tilde{Y}$, respectively. To get an analogue result for an approximation of these singular values, we need to get an operator corresponding to the matrix

$$\begin{pmatrix} 0 & Y^T \\ Y & 0 \end{pmatrix} \in \mathbb{R}^{m+N, m+N}$$

in the finite-dimensional case.

We make use of the special structure of the range of Y, $\tilde{Y}$ by considering the finite-dimensional space

$$H_d := \operatorname{span}\{e_1^d, \ldots, e_d^d\}$$

with orthonormal basis $\{e_i^d\}_{i \leq d}$, $d \leq 2m$, such that

$$H_d = \operatorname{span}\{y_1, \ldots, y_m, \tilde{y}_1, \ldots, y_m\}.$$

We write the snapshots as linear combinations of the orthonormal basis by

$$y_i = \sum_{j=1}^{d} \alpha_{ij} e_j^d, \quad i = 1, \ldots, m$$

and collect the linear factors in $A = (\alpha_{ij})_{ij} \in \mathbb{R}^{m,d}$. For the snapshots $\tilde{y}_i$ we do the same and get a corresponding matrix $\tilde{A} = (\tilde{\alpha}_{ij})_{ij} \in \mathbb{R}^{m,d}$. We identify H_d as a finite-dimensional Hilbert space with the $\mathbb{R}^d$ via the isometry $J : H_d \to \mathbb{R}^d$ defined by

$$Je_i^d = e_i, \quad i = 1, \ldots, m$$

where $\{e_i\}_{i \leq d}$ is the standard basis in $\mathbb{R}^d$. An easy calculation shows that the operator $Y : \mathbb{R}^m \to H_d$ can be written as

$$Y(v) = \frac{1}{\sqrt{m}} J^{-1}(A^T v)$$

and the adjoint as

$$Y^*(\varphi) = \frac{1}{\sqrt{m}} A J(\varphi).$$

Hence the operator $Y^*Y : \mathbb{R}^m \to \mathbb{R}^m$ is simply given by the matrix $\frac{1}{m} A A^T \in \mathbb{R}^{m,m}$. By definition (see Theorem 2.2) the singular values of Y coincide with the singular values of $\frac{1}{\sqrt{m}} A^T$. Since the same relationship holds for $\tilde{Y}$ and the matrix $\frac{1}{\sqrt{m}} \tilde{A}^T$ we can apply the finite-dimensional perturbation theory of Theorem 2.7 using the matrix

$$\begin{pmatrix} 0 & A \\ A^T & 0 \end{pmatrix} \in \mathbb{R}^{m+d,m+d}$$

to get the result

$$\sum_{i=1}^{m} (\sigma_i - \tilde{\sigma}_i)^2 \leq \frac{1}{m} \sum_{i=1}^{m} \|a_i - \tilde{a}_i\|_2^2$$

where $A^T = \mathrm{col}(a_1, \ldots, a_m)$. We get the same result as in (2.8) if we observe that

$$\begin{aligned}
\|a_i - \tilde{a}_i\|_2 &= \|A^T e_i - \tilde{A}^T e_i\|_2 \\
&= \|J^{-1}(A^T e_i - \tilde{A}^T e_i)\|_H = \|y_i - \tilde{y}_i\|_H \quad \text{for all } i = 1, \ldots, m.
\end{aligned}$$

Hence we showed:

Corollary 2.8. *Let H be a separable Hilbert space and two collections $\{y_i\}_i$ and $\{\tilde{y}_i\}_i$ of snapshots be given in H. Then for the singular values of the corresponding operators $Y, \tilde{Y} : \mathbb{R}^m \to H$ defined in (2.3), the following holds:*

$$\sum_{i=1}^{m} (\sigma_i - \tilde{\sigma}_i)^2 \leq \frac{1}{m} \sum_{i=1}^{m} \|y_i - \tilde{y}_i\|_H. \tag{2.9}$$

Remark. *As a more general approach to the infinite-dimensional case it seems reasonable to consider the Hilbert space $H \times \mathbb{R}^m$ with scalar product*

$$\langle (\varphi, v), (\theta, w) \rangle_{H \times \mathbb{R}^m} = \langle \varphi, \psi \rangle + \langle v, w \rangle_2$$

as state space for the operator $\mathcal{Y}$. Then one can define the operator $\mathcal{Y} : H \times \mathbb{R}^m \to H \times \mathbb{R}^m$ as an analogue to $\begin{pmatrix} 0 & Y \\ Y^T & 0 \end{pmatrix}$ by setting

$$\mathcal{Y}(\varphi, v) = (Y(v), Y^*(\varphi)), \quad v \in \mathbb{R}^m, \varphi \in H. \tag{2.10}$$

It can easily be seen that $\mathcal{Y}$ is self-adjoint and eigenpairs of $\mathcal{Y}$ correspond to singular triples of Y. There is a similar perturbation result to Theorem 2.7 for self-adjoint operators in general Hilbert spaces. Given self-adjoint operators S, T in H it states that the Hausdorff *distance of the spectra* $\mathrm{dist}(\Sigma(S), \Sigma(T))$ *can be estimated by the operator norm of the difference* $\|S - T\|_H$ *(see [Kat76], Theorem V.4.10). Applied to* $\mathcal{Y}$ *and* $\tilde{\mathcal{Y}}$*, defined analogously for the snapshots* $\tilde{y}_i$*, we get the estimate*

$$\mathrm{dist}(\{\sigma_i\}_i, \{\tilde{\sigma}_i\}_i)^2 \leq \frac{1}{m} \sum_{i=1}^{m} \|y_i - \tilde{y}_i\|^2.$$

Nevertheless this is a weaker result than (2.9) since

$$(\mathrm{dist}(\{\sigma_i\}_i, \{\tilde{\sigma}_i\}_i))^2 = (\max(\max_i(\min_j |\sigma_i - \tilde{\sigma}_j|), \max_j(\min_i |\sigma_i - \tilde{\sigma}_j|)))^2$$
$$\leq (\max_i |\sigma_i - \tilde{\sigma}_i|)^2 = \max_i(\sigma_i - \tilde{\sigma}_i)^2$$
$$\leq \sum_{i=1}^{m}(\sigma_i - \tilde{\sigma}_i)^2$$

where the last inequality is a real inequality in general.

2.2 POD convergence theory on finite time intervals

We give a brief overview of the existing convergence theory of POD methods for parabolic problems developed by Kunisch, Volkwein et al. (see [KV01], [KV02]).

We define a parabolic initial boundary value problem by

$$\frac{\partial u}{\partial t} + Au = f(u) \quad \text{in } \Omega$$
$$u = 0 \quad \text{on } \partial\Omega \qquad (2.11)$$
$$u(x, 0) = u_0(x), \quad x \in \Omega$$

where $A : V \to H$ is a linear positive operator, $f \in C(V, H)$ with real separable Hilbert spaces V, H defining a so-called *Gelfand triple*, i.e. V is dense in H with continuous injection and, by identifying H and its dual H^*, we get dense embeddings

$$V \subset H = H^* \subset V^*.$$

If we choose $Au = -\Delta u$ and f as a proper nonlinearity, we get a reaction-diffusion system. For example we get the so-called *Chafee-Infante problem* on $\Omega = (0, 1)$ with $V = H_0^1(\Omega)$, $H = L_2(\Omega)$ with the following nonlinearity

$$\frac{\partial u}{\partial t} - \Delta u = \lambda(u^3 - u) \quad \text{in } \Omega$$
$$u(0, t) = u(1, t) = 0, \quad t \geq 0 \qquad (2.12)$$
$$u(x, 0) = u_0(x), \qquad x \in [0, 1].$$

We will have a closer look at this well-analyzed parabolic system later on in Chapter 6 to test our algorithms on it.

The PDE in (2.11) can be transformed in variational formulation by multiplication from the right with $\varphi \in V$. Assuming that A induces a V-elliptic continuous bilinear form $a : V \times V \to \mathbb{R}$, we see that for given $T > 0$ the problem transforms to

$$\frac{d}{dt}(u(t), \varphi)_H + a(u(t), \varphi) = (f(u(t)), \varphi)_h \quad \text{for all } \varphi \in V, t \in (0, T)$$
$$(u(0), \chi)_H = (u_0, \chi)_H \quad \text{for all } \chi \in H. \qquad (2.13)$$

Under proper conditions for the nonlinearity f there is a unique and continuous solution to (2.13) on a finite time interval (0,T), the so-called weak solution of the PDE (2.11). We denote this solution by $u(t) = u(t; T, u_0) \in C([0, T], V)$.

Volkwein and Kunisch consider a POD basis derived from such a trajectory $u(t)$ at given time steps $t_j = j\Delta t$, $j = 1, \ldots, m$. To achieve a better error constant, they also include the corresponding finite difference quotients

$$\bar{\partial} u(t_k) = \frac{u(t_k) - u(t_{k-1})}{\Delta t}$$

to obtain a collection of snapshots

$$y_j = u(t_{j-1}), \, j = 1, \ldots, m+1,$$
$$y_{m+1+j} = \bar{\partial} u(t_{j-1}), \, j = 1, \ldots, m+1.$$

For these snapshots, let the POD basis of rank ℓ be given by $\{\psi_1, \ldots, \psi_\ell\}$ and denote the ℓ-dimensional POD space by

$$V^\ell := \text{span}\{\psi_1, \ldots, \psi_\ell\}.$$

Now we can set up the reduced-order system corresponding to (2.11) using the backward Euler-Galerkin scheme

$$(\bar{\partial} U_k, \psi)_H + a(U_k, \psi) = (f(U_k), \psi)_H \quad \text{for all } \psi \in V^\ell, \tag{2.14}$$
$$(U_0, \psi)_H = (u_0, \psi)_H \quad \text{for all } \psi \in V^\ell.$$

Here, $\bar{\partial} U_k$ denotes the analogue to $\bar{\partial} u(t_k)$ for the Euler sequence in $(U_k)_k$ in V^ℓ:

$$\bar{\partial} U_k = \frac{U_k - U_{k-1}}{\Delta t}.$$

Kunisch and Volkwein proved the following relation between the exact weak solution and the POD solution for a given initial value $u_0 \in H$:

Theorem 2.9. *Assume that (2.13) has a unique solution $u \in C([0,T],V)$ with $u \in W^{2,2}([0,T];H)$ and that $\{U_k\}_{k=0}^m$ is the unique solution to (2.14) satisfying*

$$\max_{0 \leq k \leq m} \|U_k\|_H \leq \tilde{C}$$

for a constant $\tilde{C} > 0$ independent of m. If f is locally Lipschitz on H and Δt sufficiently small, then there exists a constant $C > 0$ independent of ℓ, m such that

$$\frac{1}{m} \sum_{k=1}^m \|u(t_k) - U_k\|_H^2 \leq C \left(\|u_0 - P^\ell u_0\|_H^2 + \sum_{\ell+1}^d \sigma_k^2 + (\Delta t)^2 \right).$$

Here, the Ritz projector $P^\ell : H \to V^\ell$ is defined by

$$a(P^\ell u, \psi) = a(u, \psi) \quad \text{for all } \psi \in V^\ell$$

and σ_k are the singular values of Y defined by (2.3).

2.3 Long-time behavior of POD solutions

To our knowledge, error estimates concerning POD modes exist only for finite time intervals. In Section 2.2 we have sketched an error estimate for the approximation of single trajectories in finite time intervals by POD modes.

We want to use the model reduction via POD modes in set-oriented methods introduced in Chapter 1. These algorithms approximate attractors and invariant measures, i.e. long-time properties of underlying dynamical systems.

This motivates an analysis of how aspects of the long-time behavior transfer to reduced order systems if we use POD-based model reduction. A detailed description of every possible dynamical behavior seems impossible. Hence we restrict ourselves to some typical long-time properties to get a feeling for possibilities and limits of the POD methods for the analysis of long-time properties of dynamical systems.

A central error bound in the convergence theory of Volkwein is given by the behavior of the remaining singular values

$$\sum_{k=\ell+1}^{d} \lambda_k$$

of the operator Y given by the snapshots. We will have a closer look at the time dependence of this error bound in the following cases. For simplicity we restrict ourselves to the finite-dimensional case. We consider the ordinary differential equation

$$u_t = f(u), \quad f : \mathbb{R}^N \to \mathbb{R}^N$$
$$u(0) = u_0. \tag{2.15}$$

Wherever possible, we will make generalizations to the case of an infinite dimensional state space.

2.3.1 Asymptotically stable fixed point

We assume that there exists an asymptotically stable fixed point in ODE (2.15). We consider snapshots based on a trajectory converging to that fixed point. As we expect, the resulting POD basis converges to the same fixed point. In detail, the following holds:

Theorem 2.10. Let $\bar{u}$ be an asymptotically stable fixed point of (2.15), i.e. $f(\bar{u}) = 0$ and

$$\sigma(Df(\bar{u})) \subset \mathbb{C}_- := \{z \in \mathbb{C}; \Re z < 0\}. \tag{2.16}$$

Let $u(t)$ be a trajectory of (2.15) with initial value u_0 near $\bar{u}$ such that

$$\|u(t) - \bar{u}\|_2 \le e^{-\alpha t}\|u_0 - \bar{u}\|_2 \quad \text{for all } t > 0. \tag{2.17}$$

and some $\alpha > 0$. Then the POD basis of rank 1 for the snapshots

$$y_j = u(t_j) = u(jT), \quad j = 1, \dots, m$$

with stepsize $T \ge T_0 > 0$ is given by some $\{w_1\}$ where the error of the POD method (see Corollary 2.5) is given by

$$E_{\text{pod}}(\{w_1\}) \le C_1 \frac{e^{-\alpha 2T}}{m} \tag{2.18}$$

and the angle between the POD mode and the fixed point satisfies

$$|\sin(\angle(w_1, \bar{u}))| \le C_2 \frac{e^{-\alpha T}}{\sqrt{m}} \tag{2.19}$$

with C_1, C_2 independent of T, m.

Remarks. • *For the definition of the canonical angle $\angle(V, W)$ between the subspaces given by $V \in \mathbb{R}^{n,m}$ and $W \in \mathbb{R}^{n,m}$ see appendix A.2 where aspects of the perturbation theory concerning singular value decompositions are presented.*

• *The existence of a trajectory with (2.17) follows from the assumption (2.16). Therefore consider the linearization of (2.15) at $\bar{u}$. With $v = u - \bar{u}$ the equation (2.15) transforms to*

$$\begin{aligned} v_t(t) = (u(t) - \bar{u})_t &= f(v(t) + \bar{u}) \\ &= Df(\bar{u})v(t) + O(|v(t)|^2) \end{aligned} \tag{2.20}$$
$$v(0) = u_0 - \bar{u}$$

Observe that the corresponding linear system

$$\begin{aligned} w_t &= Df(\bar{u})w \\ w(0) &= u_0 - \bar{u} \end{aligned} \tag{2.21}$$

is solved by $w(t) = e^{At}(u_0 - \bar{u})$ with $A := Df(\bar{u})$ and hence satisfies

$$\|w(t)\|_2 \le e^{-\lambda t} \|u_0 - \bar{u}\|_2$$

with $\lambda = -\min(-\sigma(Df(\bar{u}))) > 0$. It follows immediately that a trajectory $u(t)$ starting at u_0 in a small neighborhood of $\bar{u}$ satisfies (2.17) and therefore $\bar{u}$ is an asymptotically stable fixed point. For details we refer to [SH96] and [Wig03].

Proof. We consider the matrix $Y = \frac{1}{\sqrt{m}} \operatorname{col}(y_1, \dots, y_m)$ and decompose it in the form

$$Y = Y_0 + E, \quad Y_0 = \frac{1}{\sqrt{m}} \operatorname{col}(\bar{u}, \dots, \bar{u}) \in \mathbb{R}^{N,m}.$$

The singular value decomposition of Y_0 is given by

$$W_0^T Y_0 V_0 = \begin{pmatrix} \|\bar{u}\|_2 & 0 \\ 0 & 0 \end{pmatrix}$$

where $W_0 = \operatorname{col}(w_1, \dots, w_N)$ with

$$w_1 = \frac{\bar{u}}{\|\bar{u}\|_2}.$$

The singular values $\{\sigma_i^0\}_i$ of Y_0 are given by

$$\sigma_1^0 = \|\bar{u}\|_2, \ \sigma_i^0 = 0, \quad i \ge 2.$$

Let the singular value decomposition of Y be given by

$$W^T Y V = \begin{pmatrix} \operatorname{diag}(\sigma_1, \dots, \sigma_m) \\ 0 \end{pmatrix}.$$

Then by Theorem 2.7

$$
\begin{aligned}
E_{\text{pod}}(\{w_1\}) \;&=\; \sum_{i=2}^{m} \sigma_i^2 \le (\sigma_1 - \|\bar{u}\|_2)^2 + \sum_{i=2}^{m} \sigma_i^2 \\
&=\; \sum_{i=1}^{m} (\sigma_i - \sigma_i^0)^2 \\
&\overset{\text{(Thm. 2.7)}}{\le}\; \frac{1}{m} \sum_{i=1}^{m} \|y_i - \bar{u}\|_2^2 \\
&\le\; \frac{1}{m} \sum_{i=1}^{m} \left(e^{-2\alpha T}\right)^i \|u_0 - \bar{u}\|_2^2 \\
&\le\; \frac{1}{m} e^{-\alpha 2T} \frac{1}{1 - e^{-2\alpha T_0}} \|u_0 - \bar{u}\|_2^2.
\end{aligned}
\tag{2.22}
$$

For the singular vectors recall that by Corollary A.13 the canonical angle between the singular vector w_1 of Y and $\bar{u}$ of Y_0 is bounded by

$$
|\sin \angle(w_1, \bar{u})| \le \frac{\|E\|_2}{\gamma - \|E\|_2}
$$

where the spectral gap γ in this example is given by

$$
\gamma = \sigma_1^0 - \sigma_2^0 = \|\bar{u}\|_2.
$$

For $E = Y - Y_0 = \text{col}(y_1 - \bar{u}, \dots, y_m - \bar{u})$ we use the approximation

$$
\|E\|_2 \le \|E\|_F = \left(\frac{1}{m} \sum_{i=1}^{m} \|y_i - \bar{u}\|_2^2\right)^{1/2}.
$$

If we take m or T large enough such that

$$
\|E\|_F \le \frac{\|\bar{u}\|_2}{2}
$$

we get

$$
|\sin(\angle(w_1, \bar{u}))| \le \frac{2\|E\|_F}{\|\bar{u}\|_2} \overset{(2.22)}{\le} \frac{2C}{\sqrt{m}} \|\bar{u} - u_0\| e^{-\alpha T} \sqrt{\frac{1}{1 - e^{-2\alpha T_0}}}.
$$

$\square$

Remark. *Note that we immediately get an analogue result for the case of an infinite-dimensional state space by Theorem 2.8.*

2.3.2 Attracting invariant subspace

Now we consider a positive invariant subspace that attracts the trajectory defining the snapshots. This is an easy generalization of the case of an asymptotically stable fixed point. Roughly speaking the behavior of the POD modes transfers. Note that we have to make additional regularity assumptions for the subspace property. This is due to the fact that the POD algorithm has too many degrees of freedom if the trajectory does not exhaust the subspace.

Theorem 2.11. *Let* $\mathcal{Z} = \text{span}\{z_1, \dots, z_\ell\} \subset \mathbb{R}^N$ *be an ℓ-dimensional subspace of $\mathbb{R}^N$ that is positive invariant under (2.15) and exponentially attracts a trajectory $u(t)$:*

$$\text{dist}(u(t), \mathcal{Z}) \leq Ce^{-\alpha t}\,\text{dist}(u_0, \mathcal{Z}). \tag{2.23}$$

We choose the snapshots $y_j = u(jT)$, $j = 1, \dots, m$ along this trajectory. If we assume the trajectory to be regular in the sense that

$$\text{rank}(ZZ^T Y) = \ell \tag{2.24}$$

with $Z = \text{col}(z_1, \dots, z_\ell)$, $Y = \text{col}(y_1, \dots, y_m)$, then for the POD space $\text{span}\{w_1, \dots, w_\ell\}$ of rank ℓ, we get the estimate

$$E_{\text{pod}}(\{w_i\}_i) \leq C_1 \frac{e^{-\alpha 2T}}{m}\,\text{dist}^2(u_0, \mathcal{Z}).$$

If the trajectory is essential for the subspace $\mathcal{Z}$ in the sense that

$$\sigma_\ell(ZZ^T Y) \geq C > 0 \tag{2.25}$$

with C independent of T, m, then in addition we have

$$\|\sin(\angle(Z, W))\|_2 \leq C_2 \frac{e^{-\alpha T}}{\sqrt{m}}\,\text{dist}(u_0, \mathcal{Z}).$$

As above $C_1, C_2 > 0$ are independent of T, m.

Remark. *Note that the assumption (2.23) follows the concept of inertial manifolds. This concept is described in [Tem97] and [Rob01] for infinite dimensions.*

Proof. Consider the orthogonal projection $P_{\mathcal{Z}} : \mathbb{R}^N \to \mathbb{R}^N$ onto $\mathcal{Z}$ that can be expressed by

$$P_{\mathcal{Z}} v = ZZ^T v \in \mathcal{Z}.$$

We can express the Hausdorff distance by

$$\text{dist}^2(u(t), \mathcal{Z}) = \|u(t) - P_{\mathcal{Z}} u(t)\|_2^2 = \|(I - ZZ^T)u(t)\|_2^2.$$

As before, we collect the snapshots $y_j = u(jT)$ along the trajectory u(t) in a matrix $Y = \frac{1}{\sqrt{m}}\text{col}(y_1, \dots, y_m) \in \mathbb{R}^{N,m}$. We can decompose Y by

$$Y = Y_0 + E, \ Y_0 = ZZ^T Y, \ E = (I - ZZ^T)Y.$$

We have the following singular value decomposition of Y_0

$$W_0^T Y_0 V_0 = \begin{pmatrix} \text{diag}(\sigma_1^0, \dots, \sigma_\ell^0) & 0 \\ 0 & 0 \end{pmatrix}$$

with positive singular values $\sigma_1^0 \geq \ldots \geq \sigma_\ell^0 > 0$ by assumption (2.24). By standard perturbation theory, we get a result for the singular values $\sigma_1 \geq \ldots \geq \sigma_d > 0$ of Y:

$$
\begin{aligned}
\sum_{k=\ell+1}^{d} \sigma_k^2 &\leq \sum_{k=1}^{d} (\sigma_k - \sigma_k^0)^2 \\
&\overset{\text{(Thm. 2.7)}}{\leq} \frac{1}{m} \sum_{j=1}^{m} \|(I - ZZ^T)y_j\|_2^2 = \frac{1}{m} \sum_{j=1}^{m} \text{dist}^2(y_j, \mathcal{Z}) \\
&\leq \frac{C}{m} \text{dist}^2(u_0, \mathcal{Z}) \sum_{j=1}^{m} e^{-2jT} \\
&\leq \frac{e^{-2\alpha T}}{m} C \frac{1}{1 - e^{-2\alpha T_0}} \text{dist}^2(u_0, \mathcal{Z}).
\end{aligned}
$$

For the singular vectors observe that we need a spectral gap to get an estimate for the canonical angles. If the assumption (2.25) holds, the spectral gap is just given by $C > 0$ and we can use Corollary A.13 as before to get the result. If we take m or T large enough such that

$$\|E\|_2 \leq \|E\|_F \leq \frac{1}{2}C$$

we get

$$\|\sin(\angle(Z, W))\|_2 \leq \frac{2\|E\|_2}{C} \leq C_2 \frac{e^{-\alpha T}}{\sqrt{m}} \text{dist}(u_0, \mathcal{Z}).$$

$\square$

2.3.3 Two rates of convergence

In the next theorem we want to deal with the situation of an asymptotically stable fixed point in detail. Therefore we have a closer look at the linearized system (2.21) around a stable fixed point $\bar{u}$. Assume $A := Df(\bar{u})$ to be diagonalizable ($VAV^{-1} = D$). For the solution $w(t)$ of (2.21) with initial value $u_0 - \bar{u}$, this yields

$$
\begin{aligned}
w(t) &= e^{At}(u_0 - \bar{u}) = Ve^{Dt}V^{-1}(u_0 - \bar{u}) \\
&= V \operatorname{diag}(e^{-\lambda_1 t}, \ldots, e^{-\lambda_N t}) \underbrace{V^{-1} v_0}_{=:w_0} \\
&= \sum_{i=1}^{N} (w_0)_i e^{-\lambda_i t} v_i
\end{aligned}
$$

where $V = \text{col}(v_1, \ldots, v_N)$. Assuming a gap in the spectrum $\{\lambda_i\}_i$ we analyze the behavior of the POD modes for the linearized system.

Theorem 2.12. *Let a trajectory converge to the fixed point 0 in the following way:*

$$u(t) = e^{-\lambda_1 t} v + e^{-\lambda_2 t} w \tag{2.26}$$

where $\lambda_2 = \lambda_1 + \delta$. Then the resulting singular values controlling the POD expansion for the snapshots $y_j = u(jT)$, $j = 1, \ldots, m$, have a gap

$$\sigma_1 \in \left[\left(1 - \gamma \frac{\|w\|_2}{\|v\|_2}\right)\sigma_1^0, \left(1 + \gamma \frac{\|w\|_2}{\|v\|_2}\right)\sigma_1^0 \right], \tag{2.27}$$

$$|\sigma_2| \le \gamma \frac{\|w\|_2}{\|v\|_2}\sigma_1^0. \tag{2.28}$$

Here, $\gamma = e^{-\delta T}$ and σ_1^0 is the first singular value of

$$Y_0 = \frac{1}{\sqrt{m}} \operatorname{col}(e^{-\lambda_1 T} v, \ldots, e^{-\lambda_1 mT} v).$$

This leads to the following estimate for the angle between the first POD mode w_1 of Y and the direction of slowest attraction v

$$|\sin(\angle(v, w_1))| \le \frac{\gamma}{\frac{\|v\|_2}{\|w\|_2} - \gamma} \tag{2.29}$$

if $\frac{\|v\|_2}{\|w\|_2} > \gamma$.

Proof. As before, let $Y = \frac{1}{\sqrt{m}} \operatorname{col}(y_1, \ldots, y_m)$ be the matrix of snapshots. It can be written as a sum of two rank-1-matrices

$$Y = Y_0 + E, \quad Y_0 = \frac{1}{\sqrt{m}} v a^T, \quad E = \frac{1}{\sqrt{m}} w b^T$$

where $a = (e^{-\lambda_1 T}, \ldots, e^{-\lambda_1 mT})^T$, $b = (e^{-\lambda_1 T}, \ldots, e^{-\lambda_1 mT})^T$. With $D = \operatorname{diag}(\gamma, \ldots, \gamma^m)$, $\gamma = e^{-\delta T}$, one can also write b as $b = Da$.

The singular values $\{\sigma_i^0\}_i$ of Y_0 are zero except for

$$\sigma_1^0 = \frac{1}{\sqrt{m}} \|v\|_2 \|a\|_2.$$

The same holds for the singular values $\{\sigma_i^1\}_i$ of E with

$$\|E\|_2 = \sigma_1^1 = \frac{1}{\sqrt{m}} \|w\|_2 \|Da\|_2$$

$$\le \frac{\gamma}{\sqrt{m}} \|w\|_2 \|a\|_2 \quad = \gamma \frac{\|w\|_2}{\|v\|_2}\sigma_1^0. \tag{2.30}$$

We get the following relative error bounds for the singular values of Y:

$$|\sigma_1 - \sigma_1^0|, |\sigma_2| \le \|E\|_2 \le \gamma \frac{\|w\|_2}{\|v\|_2}\sigma_1^0$$

which leads to (2.27) and (2.28).

Considering the estimate of the POD mode w_1 note that the singular vector of Y_0 is given by $\frac{v}{\|v\|_2}$. We use Corollary A.13 to get an estimate for the angle between the first singular vector w_1 of Y and v:

$$|\sin(\angle(v, w_1))| \le \frac{\|E\|_2}{\delta_Y - \|E\|_2}$$

with spectral gap

$$\delta_Y = \sigma_1^0 - \sigma_2^0 = \sigma_1^0 = \frac{1}{\sqrt{m}}\|v\|_2\|a\|_2$$

and the norm of E already computed in (2.30). Together this yields

$$|\sin(\angle(v,w_1))| \leq \frac{\gamma\frac{\|w\|_2}{\|v\|_2}\sigma_1^0}{\sigma_1^0 - \gamma\frac{\|w\|_2}{\|v\|_2}\sigma_1^0} = \frac{\gamma}{\frac{\|v\|_2}{\|w\|_2} - \gamma}.$$

$\square$

This theorem shows that the first POD vector approximates the direction of the slower attraction quite well for a two-dimensional linear system. However, in the more realistic case of N different directions of attraction, the result is less satisfying as we will see in the next chapter.

2.3.4 Different speeds of convergence in the diagonalizable case

Thinking of the situation in the linear system (2.21), we look for a more general result of N different directions. For this, we need a norm estimate for the inverse of the Vandermonde matrix.

Theorem 2.13. *For distinct numbers $x_i \in \mathbb{C}$, $i = 1, \ldots, n$ define the Vandermonde matrix $V = \mathrm{Vand}(\{x_i\}_{i=1}^n) \in \mathbb{C}^{n,n}$ by*

$$V = \begin{pmatrix} 1 & x_1 & \cdots & x_1^{n-1} \\ 1 & x_2 & \cdots & x_2^{n-1} \\ \vdots & \vdots & & \vdots \\ 1 & x_n & \cdots & x_n^{n-1} \end{pmatrix}$$

Then we have

$$\|V^{-1}\|_1 \leq \max_{j=1,\ldots,n} \prod_{\substack{k=1 \\ k \neq j}}^n \frac{1+|x_k|}{|x_k - x_j|} \tag{2.31}$$

where $\|\cdot\|_1$ is the usual 1-norm induced by the sum norm.

$$\|A\|_1 = \max_{j=1,\ldots,n} \sum_{i=1}^n |a_{ij}|, \ A \in \mathbb{C}^{n,n}.$$

If the x_j are located on the same ray through the origin, i.e.

$$x_j = |x_j|e^{i\varphi}$$

for a fixed $\varphi \in [0, 2\pi)$, then (2.31) is actually an equality.

Proof. A complete proof is given in [Gau62]. In the special case of nonnegative real numbers $x_1 > \ldots > x_n > 0$ the proof is quite simple. Since we only need this case we sketch the proof here.

Let $V = \mathrm{Vand}(\{x_i\}_{i=1}^n) \in \mathbb{R}^{n,n}$ be the corresponding Vandermonde matrix to x_i as above. Since

$$\det V = \prod_{1 \le i < j \le n} (x_i - x_j) \ne 0$$

the inverse $W = (w_{ij})_{ij} \in \mathbb{R}^{n,n}$ of V exists. For each $j = 1,\ldots,n$, the j-th column $w_j = (w_{1j},\ldots,w_{nj})^T$ of W solves

$$V w_j = e_j \tag{2.32}$$

where e_j is the j-th canonical basis vector of $\mathbb{R}^N$.

Observe that the polynomial

$$x \mapsto \sum_{i=1}^n a_i x^{i-1}$$

is the unique interpolation polynomial to the data (x_j, b_j), $j = 1,\ldots,n$, if and only if $a \in \mathbb{C}^n$ solves $Va = b$. Thus, (2.32) implies for $p_{w_j}(x) = \sum_{k=1}^n w_{kj} x^{k-1}$:

$$p_{w_j}(x_i) = \delta_{ij}, \quad i,j = 1,\ldots,n$$

which means that p_{w_j} is simply the j-th Lagrange polynomial to the grid points x_i:

$$\sum_{k=1}^n w_{kj} x^{k-1} = L_j(x) = \prod_{\substack{k=1 \\ k \ne j}}^n \frac{x - x_k}{x_j - x_k}. \tag{2.33}$$

By induction it is easy to see that the coefficients of the Lagrange polynomial to ordered grid points satisfy

$$\mathrm{sign}(w_{kj}) = (-1)^{k+j}.$$

Using (2.33) with $x = -1$ we get

$$(-1)^j \sum_{k=1}^n |w_{kj}| = \sum_{k=1}^n w_{kj}(-1)^k$$

$$= \prod_{\substack{k=1 \\ k \ne j}}^n \frac{-1 - x_k}{x_j - x_k}$$

$$= (-1)^{n-1} \prod_{\substack{k=1 \\ k \ne j}}^n \frac{1 + x_k}{x_j - x_k}$$

$$= \frac{(-1)^{n-1}}{(-1)^{n-1-j}} \prod_{\substack{k=1 \\ k \ne j}}^n \frac{1 + x_k}{|x_j - x_k|} = (-1)^j \prod_{\substack{k=1 \\ k \ne j}}^n \frac{1 + x_k}{|x_j - x_k|}.$$

Multiplying with $(-1)^j$ and maximizing over j gives (2.31). $\qquad\square$

Theorem 2.14. *Let a trajectory of (2.21) be given by*

$$u(t) = \sum_{j=1}^N e^{-\lambda_j t} v_j \tag{2.34}$$

where $\{v_1, \ldots, v_N\}$ *are linearly independent and*

$$0 < \lambda_1 < \ldots < \lambda_\ell < \lambda_\ell + \delta = \lambda_{\ell+1} \leq \ldots \leq \lambda_N. \tag{2.35}$$

Collect the directions v_i *in matrices* $V_0 = \mathrm{col}(v_1, \ldots, v_\ell)$ *and* $V_1 = \mathrm{col}(v_{\ell+1}, \ldots, v_N)$. *Consider snapshots* $y_i = u(iT)$, $i = 1, \ldots, m$, *to a given stepsize* $T \geq 1/\delta$. *Then the corresponding POD basis of rank* ℓ *given by* $W_\ell = \mathrm{col}(w_1, \ldots, w_\ell)$ *satisfies*

$$\| \sin(\angle(V_0, W_\ell)) \|_2 \leq \frac{\gamma}{M - \gamma} \tag{2.36}$$

if $M > \gamma := e^{-\delta T}$. *Here,* $M > 0$ *is given by*

$$M = \begin{cases} \frac{\|v_1\|_2}{\sqrt{2(N-1)}\sigma_1(V_1)}, & \ell = 1 \\ \sqrt{\frac{\ell}{2(N-\ell)}} \left(\frac{\mathrm{gap}_\alpha}{2}\right)^\ell \frac{\sigma_\ell(V_0)}{\sigma_1(V_1)}, & \ell \geq 2 \end{cases} \tag{2.37}$$

with $\alpha_i = e^{-\lambda_i T}$ *and* $\mathrm{gap}_\alpha = \min\limits_{1 \leq i < j \leq \ell} |\alpha_i - \alpha_j|$.

If the eigenvalues are well-separated, i.e.

$$\lambda_j - \lambda_{j-1} \geq \delta \quad \text{for all } j = \ell + 1, \ldots, N \tag{2.38}$$

and $T \geq 2/\delta$, *then the bound* (2.36) *improves by choosing a larger* M *given by*

$$M = \begin{cases} \frac{\|v_1\|_2}{\sqrt{2}\sigma_1(V_1)}, & \ell = 1 \\ \sqrt{\frac{\ell}{2}} \left(\frac{\mathrm{gap}_\alpha}{2}\right)^\ell \frac{\sigma_\ell(V_0)}{\sigma_1(V_1)}, & \ell \geq 2. \end{cases} \tag{2.39}$$

Proof. As before we define the matrix of snapshots $Y = \frac{1}{\sqrt{m}} \mathrm{col}(y_1, \ldots, y_m)$. Since we have

$$y_i = u(iT) = \sum_{j=1}^{N} e^{-\lambda_j iT} v_j, \quad i = 1, \ldots, m,$$

we can write Y as

$$Y = \frac{1}{\sqrt{m}} V A^T, \ A = (\alpha_j^i)_{ij} \in \mathbb{R}^{m,N}.$$

As before, we describe the POD space of rank ℓ as a perturbation of the first ℓ eigendirections v_i, $i = 1, \ldots, \ell$. The POD space $W_\ell = \mathrm{col}(w_1, \ldots, w_\ell)$ is given by the singular value decomposition of Y:

$$W^T Y U = \begin{pmatrix} \mathrm{diag}(\sigma_1, \ldots, \sigma_N) & 0 \end{pmatrix}$$

with $W = \mathrm{col}(w_1, \ldots, w_N) \in \mathbb{R}^{N,N}$, $U \in \mathbb{R}^{m,m}$ orthogonal, $\sigma_1 \geq \ldots \geq \sigma_N \geq 0$.

We decompose Y as

$$Y = Y_0 + E, \ Y_0 = \frac{1}{\sqrt{m}} V_0 A_0^T, \ E = \frac{1}{\sqrt{m}} V_1 A_1^T$$

with $A = \begin{pmatrix} A_0 & A_1 \end{pmatrix}$, $V = \begin{pmatrix} V_0 & V_1 \end{pmatrix}$, $A_0 = (\alpha_j^i)_{ij} \in \mathbb{R}^{m,\ell}$, $V_0 = \mathrm{col}(v_1, \ldots, v_\ell) \in \mathbb{R}^{N,\ell}$. The singular value decomposition of the rank-ℓ matrix Y_0 is given by

$$W_0^T Y_0 U_0 = D := \begin{pmatrix} \mathrm{diag}(\sigma_1^0, \ldots, \sigma_\ell^0) & 0 \\ 0 & 0 \end{pmatrix} \in \mathbb{R}^{m,N} \tag{2.40}$$

with $W_0 = \mathrm{col}(w_1^0, \ldots, w_N^0) \in \mathbb{R}^{N,N}$, $U_0 \in \mathbb{R}^{m,m}$ orthogonal and $\sigma_1^0 \geq \ldots \geq \sigma_\ell^0 > 0$. Since $Y_0 = \frac{1}{\sqrt{m}} V_0 A_0^T$, we get $R(Y_0) = R(V_0)$. On the other hand by (2.40) we have

$$Y_0 U_0 = W_0 D_0 = \mathrm{col}(\sigma_1 w_1^0, \ldots, \sigma_\ell w_\ell^0, 0, \ldots, 0)$$

and hence $R(V_0) = R(Y_0) = \mathrm{span}\{w_1^0, \ldots, w_\ell^0\}$. As before, we can use Corollary A.13 to get an estimate for the angle between the POD basis given by $W_\ell \in \mathbb{R}^{N,\ell}$ and the eigendirections collected in $V_0 \in \mathbb{R}^{N,\ell}$:

$$\| \sin(\angle(V_0, W_\ell)) \|_2 \leq \frac{\|E\|_2}{\delta_Y - \|E\|_2}$$

where $\delta_Y = \sigma_\ell^0 - \sigma_{\ell+1}^0 = \sigma_\ell^0 = \sigma_\ell(Y_0)$.

In this formulation, we can estimate the canonical angle between the eigendirections and the POD modes in terms of singular values of Y_0 and E:

$$\| \sin(\angle(V_0, W_\ell)) \|_2 \leq \frac{\sigma_1(E)}{\sigma_\ell(Y_0) - \sigma_1(E)}.$$

We will discuss these values in the following.

1. $\sigma_\ell(Y_0)$: We use the characterization $Y_0 = V_0 A_0^T$ to get a lower bound for the smallest nonzero singular value $\sigma_\ell(Y_0)$. We use Theorem A.7 to get

$$\sigma_\ell(Y_0) \geq \frac{1}{\sqrt{m}} \sigma_\ell(V_0) \sigma_\ell(A_0).$$

For $\ell = 1$, the matrix $A_0 \in \mathbb{R}^{m,\ell}$ is given by a column vector such that

$$\sigma_\ell(A_0) = \sigma_1(A_0) = \|A_0\|_2 = \|(\alpha_1, \alpha_1^2, \ldots, \alpha_1^m)^T\|_2 \geq \alpha_1$$

and with $\sigma_\ell(V_0) = \sigma_1(V_0) = \|v_1\|_2$, we get

$$\sigma_\ell(Y_0) \geq \frac{1}{\sqrt{m}} \alpha_1 \|v_1\|_2. \tag{2.41}$$

For $\ell \geq 2$ we decompose $A_0 = \begin{pmatrix} A_0^{(1)} \\ A_0^{(2)} \end{pmatrix}$, $A_0^{(1)} \in \mathbb{R}^{\ell,\ell}$, and get a lower bound by Corollary A.6:

$$\sigma_\ell(A_0) = \sigma_\ell(A_0^T) = \sigma_\ell\left(\left(A_0^{(1)^T} \quad A_0^{(2)^T} \right) \right) \geq \sigma_\ell(A_0^{(1)}).$$

$A_0^{(1)} \in \mathbb{R}^{\ell,\ell}$ can be expressed as

$$A_0^{(1)^T} = V_\alpha \, \mathrm{diag}(\alpha_1, \ldots, \alpha_\ell)$$

where $V_\alpha = \mathrm{Vand}(\{\alpha_j\}_{j=1}^\ell)$ is the Vandermonde matrix to the numbers α_i as defined in Theorem 2.13. Together with Theorem A.7 we get

$$\sigma_\ell(A_0) \geq \sigma_\ell(A_0^{(1)}) \geq \alpha_\ell \sigma_\ell(V_\alpha) = \alpha_\ell \sigma_1(V_\alpha^{-1})^{-1} = \alpha_\ell \|V_\alpha^{-1}\|_2^{-1} \geq \alpha_\ell \sqrt{\ell} \|V_\alpha^{-1}\|_1^{-1}$$

$$= \alpha_\ell \sqrt{\ell} \left(\max_{j=1,\ldots,\ell} \prod_{\substack{k=1 \\ k \neq j}}^\ell \frac{1 + \alpha_k}{|\alpha_k - \alpha_j|} \right)^{-1} \geq \alpha_\ell \sqrt{\ell} \left(\frac{\mathrm{gap}_\alpha}{2} \right)^\ell$$

since $1 + \alpha_k \leq 2$ and by definition $|\alpha_k - \alpha_j| \geq \mathrm{gap}_\alpha$ for all $k, j \in \{1, \ldots, \ell\}$, $k \neq j$. Another application of Theorem A.7 leads to

$$\sigma_\ell(Y_0) = \sigma_\ell(\frac{1}{\sqrt{m}} V_0 A_0^T) \geq \alpha_\ell \frac{\sqrt{\ell}}{\sqrt{m}} \left(\frac{\mathrm{gap}_\alpha}{2}\right)^\ell \sigma_\ell(V_0). \tag{2.42}$$

2. $\|E\|_2 = \sigma_1(E)$: To approximate the first singular vector of $E = \frac{1}{\sqrt{m}} V_1 A_1^T$ we use the product formula for singular values in Theorem A.7 to get

$$\sigma_1(E) = \sigma_1(\frac{1}{\sqrt{m}} V_1 A_1^T) \leq \frac{1}{\sqrt{m}} \sigma_1(V_1) \sigma_1(A_1^T).$$

For the largest singular value of A_1, recall the structure of the matrix

$$A_1 = (\alpha_j^i)_{ij} \in \mathbb{R}^{m, N-\ell}$$

with $\alpha_j = e^{-\lambda_j T} \in (0, 1)$, $j = \ell + 1, \ldots, N$ ordered by $\alpha_{\ell+1} \geq \ldots \geq \alpha_N$. We use the well-known matrix norm inequality $\|B\|_2^2 \leq \|B\|_1 \|B\|_\infty$ and get

$$\sigma_1(A_1)^2 = \|A_1\|_2^2 \leq \|A_1\|_1 \|A_1\|_\infty \tag{2.43}$$

$$= \left(\max_{j=\ell+1,\ldots,N} \sum_{i=1}^m \alpha_j^i\right) \left(\max_{i=1,\ldots,m} \sum_{j=\ell+1}^N \alpha_j^i\right)$$

$$= \left(\sum_{i=1}^m \alpha_{\ell+1}^i\right) \left(\sum_{j=\ell+1}^N \alpha_j\right). \tag{2.44}$$

By assumption γ is bounded by

$$\gamma = e^{-\delta T} < 2^{-\delta T} \leq 1/2$$

and the first sum can be approximated by

$$\sum_{i=1}^m \alpha_{\ell+1}^i = \alpha_{\ell+1} \frac{1 - \alpha_{\ell+1}^m}{1 - \alpha_{\ell+1}} \leq \alpha_{\ell+1} \frac{1}{1 - \alpha_{\ell+1}}$$

$$= \gamma \alpha_\ell (1 - \gamma \alpha_\ell)^{-1} \leq \gamma \alpha_\ell (1 - \gamma)^{-1} \tag{2.45}$$

$$\leq 2\gamma \alpha_\ell. \tag{2.46}$$

For the second sum in (2.44) a weak estimate is always given by the monotonicity of α_j, $j = \ell + 1, \ldots, N$:

$$\sum_{j=\ell+1}^N \alpha_j \leq \sum_{j=\ell+1}^N \alpha_{\ell+1} = (N - \ell)\alpha_{\ell+1} = (N - \ell)\gamma \alpha_\ell. \tag{2.47}$$

Combining (2.46), (2.47) and (2.44), we get the following bound for $\sigma_1(E)$:

$$\sigma_1(E) \leq \frac{1}{\sqrt{m}} \sigma_1(A_1)\sigma_1(V_1) \leq \sqrt{\frac{2(N - \ell)}{m}} \gamma \, \alpha_\ell \, \sigma_1(V_1). \tag{2.48}$$

If we have $\lambda_j - \lambda_{j-1} \geq \delta$, $j = \ell+1, \ldots, N$, we get a bound for the second sum by

$$
\sum_{j=\ell+1}^{N} \alpha_j = \alpha_{\ell+1} + \sum_{j=\ell+2}^{N} e^{-\lambda_j T} = \alpha_{\ell+1} + \sum_{j=\ell+2}^{N} e^{-\frac{\lambda_j}{\delta}\delta T} \cdot 1
$$

$$
\leq \alpha_{\ell+1} + \sum_{j=\ell+2}^{N} e^{-\frac{\lambda_j}{\delta}\delta T} \left(\frac{\lambda_j}{\delta} - \frac{\lambda_{j-1}}{\delta} \right)
$$

$$
\leq \alpha_{\ell+1} + \int_{\frac{\lambda_{\ell+1}}{\delta}}^{\frac{\lambda_N}{\delta}} e^{-s\delta T} \, ds = \alpha_{\ell+1} + \frac{1}{\delta T} \left(e^{-\lambda_{\ell+1}T} - e^{\lambda_N T} \right)
$$

$$
\leq \alpha_{\ell+1} + \frac{\alpha_{\ell+1}}{\delta T} = \gamma \alpha_\ell \left(1 + \frac{1}{\delta T} \right). \tag{2.49}
$$

With the assumption $\delta T \geq 2$ and by combining (2.45), (2.49) and (2.44), we get

$$
\sigma_1(A_1)^2 \leq \gamma\alpha_\ell(1-\gamma)^{-1}\gamma\alpha_\ell \left(1 + \frac{1}{\delta T} \right) = \gamma^2\alpha_\ell^2(1 - e^{-\delta T})^{-1} \left(1 + \frac{1}{\delta T} \right)
$$

$$
\leq \gamma^2\alpha_\ell^2(1 - 2^{-2})^{-1} \left(1 + \frac{1}{2} \right) = 2\gamma^2\alpha_\ell^2. \tag{2.50}
$$

By Theorem A.7, the product formula for singular values, we get

$$
\sigma_1(E) \leq \frac{1}{\sqrt{m}}\sigma_1(A_1)\sigma_1(V_1) \leq \sqrt{\frac{2}{m}}\gamma\,\alpha_\ell\,\sigma_1(V_1). \tag{2.51}
$$

Together for $\ell = 1$, we get

$$
\|\sin(\angle(v_1, w_1))\|_2 \quad \leq \quad \frac{\sigma_1(E)}{\sigma_1(Y_0) - \sigma_1(E)}
$$

$$
\overset{\substack{(2.41) \\ (2.48)}}{\leq} \frac{\sqrt{2(N-1)}\sigma_1(V_1)\gamma}{\|v_1\|_2 - \sqrt{2(N-1)}\sigma_1(V_1)\gamma} = \frac{\gamma}{\frac{\|v_1\|_2}{\sqrt{2(N-1)}\sigma_1(V_1)} - \gamma}
$$

provided the right hand side is positive.

For $\ell \geq 2$ we get

$$
\|\sin(\angle(V_0, W_\ell))\|_2 \quad \leq \quad \frac{\sigma_1(E)}{\sigma_\ell(Y_0) - \sigma_1(E)}
$$

$$
\overset{\substack{(2.42) \\ (2.48)}}{\leq} \frac{\sqrt{2(N-\ell)}\sigma_1(V_1)\gamma}{\sqrt{\ell}\left(\frac{\text{gap}_\alpha}{2}\right)^\ell \sigma_\ell(V_0) - \sqrt{2(N-\ell)}\sigma_1(V_1)\gamma}
$$

$$
= \frac{\gamma}{\sqrt{\frac{\ell}{2(N-\ell)}}\left(\frac{\text{gap}_\alpha}{2}\right)^\ell \frac{\sigma_\ell(V_0)}{\sigma_1(V_1)} - \gamma}
$$

provided the right hand side is positive. Similar bounds (but independent on N) hold for the case of separated eigenvalues as noted in (2.39). In that case we use (2.51) instead of (2.48). $\qquad\square$

Remark. *Since no assumptions are reasonable for the directions of attractions it does not make sense to derive further bounds on $\sigma_\ell(V_0)$ and $\sigma_1(V_1)$.*

The 2-speed case analyzed in Theorem 2.12 satisfies the assumptions of Theorem 2.14 with $\ell = 1$, $N = 2$. Hence, equation (2.36) should give the same estimate as (2.29). Indeed, this is the case except for the factor $\sqrt{2}$:

$$\angle(v, w_1) \overset{(2.36)}{\leq} \frac{\gamma}{M - \gamma} \overset{(2.39)}{\leq} \frac{\gamma}{\frac{\|v\|_2}{\sqrt{2}\,\|w_1\|_2} - \gamma}.$$

Observe that $\gamma = \gamma(T)$ is antitone in T. Hence at least for fixed dimensions of the state space, the right hand side tends to zero if the time interval T between the snapshots is increasing—as in the 2-speed case.

However, in the case of non-separated eigenvalues the right hand side depends on the state space dimension N and the error bound gets worse for increasing dimension. We cannot expect a better result, i.e. a result independent of the state space dimension, as we will see in the next theorem. There we consider a special situation satisfying the assumptions of Theorem 2.14. We will illustrate this 'worst-case scenario' later on in an example.

Fortunately, in most numerical applications the eigenvalues decay very fast, so that the stronger assumption (2.38) holds. Additionally, the higher modes v_j, $j \gg 1$, usually also decay in norm. We will see in a second example considering a linear parabolic equation that this leads to better approximation results as indicated by the second part of Theorem 2.14.

Proposition 2.15. *Let a trajectory of (2.21) be given by*

$$u(t) = \sum_{j=1}^{N} e^{-\lambda_j t}\, v_j \tag{2.52}$$

with $V = \mathrm{col}(v_1, \ldots, v_N) \in \mathbb{R}^{N,N}$ orthogonal and eigenvalues

$$0 < \lambda_1 < \lambda_2 = \lambda_1 + \delta = \lambda_3 = \ldots = \lambda_N. \tag{2.53}$$

As before, we take snapshots along the trajectory denoted by $y_i = u(iT)$, $i = 1, \ldots, m$. Let the POD basis of rank 1 be given by the vector $w_1 \in \mathbb{R}^N$. Then the angle between v_1 and w_1 increases at least with the square root of the space dimension N:

$$\left| \frac{\pi}{2} - \angle(v_1, w_1) \right| = O(N^{-1/2}). \tag{2.54}$$

Proof. We write the matrix of snapshots again as

$$Y = \frac{1}{\sqrt{m}} V A^T, \; A = (\alpha_j^i)_{ij} \in \mathbb{R}^{m,N}$$

with $\alpha_j = e^{-\lambda_j T}$. In this special case A is a rank-2 matrix of the following form

$$A = \mathrm{col}(a, Da, \ldots, Da)$$

where $D = \mathrm{diag}(\gamma, \ldots, \gamma^m)$, $\gamma = e^{-\delta T}$, and $a = (\alpha_1, \ldots, \alpha_1^m)$.

Observe that, since V is orthogonal, the following holds for the first singular vector w_1 of Y:

Let the singular value decomposition of A^T be given by

$$W_A^T A^T U_A = D_A = \text{diag}(\sigma_1, \sigma_2, 0, \ldots, 0)$$

with $W_A \in \mathbb{R}^{N,N}$, $U_A \in \mathbb{R}^{m,m}$. Now, since V is orthogonal the singular value decomposition of $Y = \frac{1}{\sqrt{m}} V A^T$ can be expressed via the decomposition of A^T:

$$(VW_A)^T Y U_A = \frac{1}{\sqrt{m}} D_A.$$

Hence the first left singular vector w_1 of Y is just given by

$$w_1 = V w_A$$

where w_A is the first column of W_A. Now we consider A as a perturbation of the rank-1 matrix $\bar{A} = \text{col}(Da, \ldots, Da)$. An easy calculation shows that $\bar{\sigma} = \sqrt{N} \|Da\|_2$ is the nontrivial singular value of $\bar{A}^T$ with left and right singular vector

$$\bar{w} = \frac{1}{\sqrt{N}} \mathbb{1}, \quad \bar{u} = \frac{Da}{\|Da\|_2}.$$

As before $V\bar{w}$ is the first left singular vector of $\bar{Y} = \frac{1}{\sqrt{m}} V \bar{A}^T$. Observe that

$$\gamma^{-m}(Da)_i = \left(\frac{1}{\gamma}\right)^{m-i} \alpha_1^i \geq \alpha_1^i \geq (1 - \gamma_i)\alpha_1^i = (a - Da)_i \quad \text{for all } 1 \leq i \leq m. \quad (2.55)$$

For $N \geq N_0 := 4\gamma^{-2m}$ and the angle $\beta := \angle(\bar{w}, w_A)$, we get

$$
\begin{aligned}
|\sin(\angle(w_1, V\bar{w}))| = |\sin(\beta)| &\leq \frac{\|\bar{A} - A\|_2}{\bar{\sigma}_1 - \|\bar{A} - A\|_2} \\
&= \frac{\|(I - D)a\|_2}{\sqrt{N}\|Da\|_2 - \|(I - D)a\|_2} \\
&\overset{(2.55)}{\leq} \frac{\gamma^{-m}}{\sqrt{N} - \gamma^{-m}} \leq \frac{2\gamma^{-m}}{\sqrt{N}} \quad \text{for all } N \geq N_0. \quad (2.56)
\end{aligned}
$$

Applying the cosine theorem and addition theorems to

$$\angle(v_1, w_1) = \angle(Ve_1, Vw_A) = \angle(e_1, w_A),$$

we get

$$
\begin{aligned}
\left| |\cos(\angle(v_1, w_1))| - \frac{1}{\sqrt{N}} \right| &\leq \left| \cos(\angle(v_1, w_1)) - \frac{1}{\sqrt{N}} \right| = |e_1^T w_A - e_1^T \bar{w}| \\
&= |e_1^T (w_A - \bar{w})| \leq \|e_1\|_2 \|w_A - \bar{w}\|_2 \\
&= \|w_A - \bar{w}\|_2 = \sqrt{(w_A - \bar{w})^T (w_A - \bar{w})} \\
&= \sqrt{2 - 2w_A^T \bar{w}} = \sqrt{2(\cos(0) - \cos(\beta))} \\
&= \sqrt{4 \sin^2\left(\frac{\beta}{2}\right)} = 2 \sin\left(\frac{\beta}{2}\right).
\end{aligned}
$$

Since $\sin(h) = h + O(h^3)$ there exists an $N_1 > N_0$ such that the statement follows by (2.56):

$$|\frac{\pi}{2} - \sin(\angle(v_1, w_1))| = |\cos(\angle(v_1, w_1))| \leq 2\sin\left(\frac{\beta}{2}\right) + \frac{1}{N} \leq \frac{C}{\sqrt{N}} \quad \text{for all } N \geq N_1.$$

$\square$

Remark. *We note that the special type of trajectories given by (2.52) and (2.53) can also be considered as the 2-speed case of Theorem 2.12 by setting $v = v_1$ and $w = \sum_{j=2}^{N} v_j$ in (2.26). This formulation shows that the setting of Proposition 2.15 is the 'worst case' not only considering the distribution of the eigenvalues but also considering the norms of the directions v and w as above:*

$$1 = \|v\|_2 \ll \|w\|_2 = \left(\sum_{i=2}^{N} \|v_i\|_2\right)^{-\frac{1}{2}} = \sqrt{N-1}$$

for large $N \in \mathbb{N}$. In this example, a reasonable error bound as by Theorem 2.12 can only be achieved for a large integration time T. If T is chosen too small, the condition $\frac{\|v\|_2}{\|w\|_2} > \gamma$ of Theorem 2.12 is not satisfied::

$$\frac{1}{\sqrt{N-1}} = \frac{\|v\|_2}{\|w\|_2} > \gamma = e^{-\delta T} \iff T > \frac{\ln\sqrt{N-1}}{\delta}.$$

Indeed, the following experiments show that the POD method does not detect the first eigendirection in this worst-case scenario.

2.3.5 Numerical examples

1. Dependency on N

In Theorem 2.14, we have stated that the POD error bound in the diagonalizable case depends on N. In our first numerical example we analyze this dependency for the worst-case scenario treated in Proposition 2.15. For this, we consider a trajectory given by (2.52) and (2.53) with different gaps in the spectrum and plot the angle between the first POD mode w_1 and the first eigendirection v_1 against the dimension of the system.

In detail, we make a numerical test with the following data: We consider the space $\mathbb{R}^{1\,000}$ with a randomly chosen orthonormal basis $\{v_i\}_{i\leq 1\,000}$. We build the trajectory $u(t) \in \mathbb{R}^{1\,000}$ according to (2.52) for a given $N \in [2, 1\,000]$ by

$$u(t) = e^{-\lambda_1 t}v_1 + e^{-\lambda_2 t}\sum_{j=2}^{N} v_j, \quad \lambda_1 < \lambda_2.$$

We use time steps $t_i = iT$, $i = 1, \ldots, m$, with $T = 10$ and $m = 100$ for the collection of snapshots $y_i = u(t_i)$, $i = 1, \ldots, m$. We fix m and T and analyze the dependency on N treated in Proposition 2.15.

The result is given by the two plots in Figure 2.1 which confirm the theoretical result of Proposition 2.15. In the left plot, we see that the angle between the first POD mode and the first eigendirection grows up to the value $\frac{\pi}{2}$, i.e. in the end no more information about the first eigendirection is contained in the first POD mode.

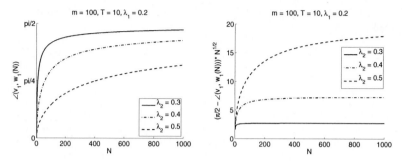

Figure 2.1: Angle between first eigen- and POD-vector in a 'worst-case' test system. Different values of λ_2 where $\lambda_1 = 0.2$ is fixed.

In the right plot, the angle is scaled according to (2.54). We see that, at least for the smaller gaps $\delta = 0.1$ and $\delta = 0.2$, the angle behaves just as the estimate of Theorem 2.15 suggests and the constant can be estimated from the plot. For $\delta = 0.3$ the limit is not reached within the observation interval $1 \leq N \leq 1\,000$ but still we can predict the convergence from the plot.

2. Linear parabolic problem

In a second example we analyze the POD behavior of a linear reaction-diffusion equation. Later on we will focus on nonlinear reaction-diffusion equations, especially the Chafee-Infante problem. This example gives a first insight into parabolic problems. It shows that the 'worst case' considered above is not typical in natural examples. We start with the scalar parabolic problem

$$
\begin{aligned}
u_t &= u_{xx} + \mu u, & t &\geq 0, x \in (0,1) \\
u(0,t) &= u(1,t) = 0, & t &\geq 0 \\
u(x,0) &= u_0(x), & x &\in [0,1]
\end{aligned}
\tag{2.57}
$$

with parameter $\mu > 0$.

Discretization by FE method As later on for the Chafee-Infante problem, we discretize this equation by the standard finite element method with linear basis functions, see [LT03] for details. We define N equally distributed grid points in the unit interval

$$
x_i = ih, \quad i = 1, \ldots, N
$$

where $h = \frac{1}{N+1}$ denotes the stepsize. Using piecewise linear basis functions $\Lambda_i : [0,1] \to \mathbb{R}$, called *hat functions* and defined by

$$
\Lambda_i(x_j) = \delta_{ij}, \quad i,j = 1, \ldots, N,
$$

we write down the weak formulation of (2.57). If we take the space

$$
V_h = \mathrm{span}\{\Lambda_i : i = 1, \ldots, N\}
$$

as ansatz and test space, the finite element solution $u_h : \mathbb{R}_+ \to V_h$ is defined by the solution of

$$(\frac{d}{dt}u_h(t), \Lambda_j)_2 + a(u_h(t), \Lambda_j) = \mu(u_h(t), \Lambda_j)_2, \quad 1 \le j \le N \tag{2.58}$$

$$u_h(0) = u_{h,0}.$$

Here, the L_2-product $(\cdot, \cdot)_2$ and the elliptic bilinear form $a(\cdot, \cdot)$ are given by

$$(u, v)_2 = \int_0^1 u(x)\,v(x)\,dx, \quad a(u, v) = \int_0^1 u_x(x)v_x(x)\,dx. \tag{2.59}$$

Observe that Green's formula is used to obtain (2.58): The boundary conditions imply

$$\int_0^1 u_{xx}(x)\,v(x)\,dx = -\int_0^1 u_x(x)v_x(x)\,dx$$

for $u, v \in V_h$. The initial value $u_{h,0} \in V_h$ for the finite element system is usually given by a projection of the original initial function u_0. Here we choose the L^2-projection $u_{h,0} = P_{L^2}^h u_0$ defined by

$$(P_{L^2}^h u_0, \Lambda_i)_2 = (u_0, \Lambda_i)_2 \quad i = 1, \dots, N.$$

For fixed $t > 0$, $u_h(t)$ is an element of V_h. Hence it is represented by a vector $\mathbf{u}(t) \in \mathbb{R}^N$ via the representation

$$u_h(t) = \sum_{i=1}^N \mathbf{u}_i(t)\Lambda_i \in V_h, \quad t \ge 0.$$

Using this representation, (2.58) transforms to

$$M_h \mathbf{u}_t + S_h \mathbf{u} = \mu M_h \mathbf{u} \tag{2.60}$$

$$\mathbf{u}(0) = \mathbf{u}_0$$

where $M_h, S_h \in \mathbb{R}^{N,N}$ are symmetric matrices with entries $M_h = ((\Lambda_j, \Lambda_i)_2)_{ij}$ and $S_h = (a(\Lambda_j, \Lambda_i))_{ij}$. The vector $\mathbf{u}_0$ corresponds to $u_{h,0}$ as above.

Explicit formula for the FE solution

The so-called *mass matrix* M_h and the *stiffness matrix* S_h are positive definite: For $0 \ne \mathbf{v} \in \mathbb{R}^N$ with corresponding function $v = \sum_{i=1}^N v_i \Lambda_i \in V_h$ it follows that

$$\mathbf{v}^T M_h \mathbf{v} = (v, v)_2 \ge 0$$

and with the ellipticity constant $\beta > 0$ of a we get

$$\mathbf{v}^T S_h \mathbf{v} = a(v, v) \ge \beta(v, v)_2 \ge 0.$$

In the one-dimensional case with linear finite elements that is considered here, M and S are tridiagonal Toeplitz matrices (i.e. constant along their diagonals) with the following entries:

$$M_h = \frac{h}{6}M, \; M = \begin{pmatrix} 4 & 1 & & \\ 1 & \ddots & \ddots & \\ & \ddots & \ddots & 1 \\ & & 1 & 4 \end{pmatrix}, \quad S = \frac{1}{h}S, \; S = \begin{pmatrix} 2 & -1 & & \\ -1 & \ddots & \ddots & \\ & \ddots & \ddots & -1 \\ & & -1 & 2 \end{pmatrix} \tag{2.61}$$

The following lemma provides the eigendecompositions of S_h and M_h.

Lemma 2.16. *Let $A = (a_{ij})_{ij}$ be a tridiagonal symmetric Toeplitz matrix with entries*

$$a_{1,1} = a_{i,i} = \alpha, \; a_{i-1,i} = a_{i,i-1} = \beta, \; i = 2, \ldots, N.$$

Then A is diagonalizable. The eigenvalues are given by $D = \operatorname{diag}(\lambda_1, \ldots, \lambda_N) \in \mathbb{R}^{N,N}$ and the corresponding orthonormal basis of eigenvectors by $V = \operatorname{col}(v_1, \ldots, v_N) \in \mathbb{R}^{N,N}$ where

$$
\begin{aligned}
A v_k &= \lambda_k v_k, \\
\lambda_k &= \alpha + 2\beta \cos(k\pi h), \quad k = 1 \ldots, N, \\
v_{k\ell} &= \sqrt{2h} \sin(k\ell\pi h), \quad k, \ell = 1, \ldots, N.
\end{aligned}
$$

As before, $h = \frac{1}{N+1}$.

Proof. see [Ise09], Lemma 10.5. □

According to Lemma 2.16, the eigendecompositions of M and S are given by

$$M = V D_M V, \quad S = V D_S V$$

with an orthogonal symmetric matrix $V \in \mathbb{R}^{N,N}$ as defined in Lemma 2.16 and diagonal matrices $D_{S,M} = \operatorname{diag}(\lambda_1^{S,M}, \ldots, \lambda_N^{S,M})$ defined by

$$\lambda_k^M = 4 + 2\cos(k\pi h), \quad \lambda_k^S = 2 - 2\cos(k\pi h), \quad k = 1 \ldots, N.$$

Solving equation (2.60) for $\mathbf{u}_t$ leads to

$$
\begin{aligned}
\mathbf{u}_t &= (\mu I_N - \frac{6}{h^2} M^{-1} S)\mathbf{u} \\
&= (\mu I_N - \frac{6}{h^2} V D_M^{-1} V V D_S V)\mathbf{u} \\
&= V(\mu I_N - \frac{6}{h^2} D_M^{-1} D_S)V\mathbf{u}.
\end{aligned}
$$

Hence the solution of (2.60) is given by

$$
\begin{aligned}
\mathbf{u}(t) &= V e^{\mu I_N - \frac{6}{h^2} D_M^{-1} D_S} V \mathbf{u}_0 \\
&= \sum_{k=1}^{N} e^{-\lambda_k t} (V\mathbf{u}_0)_k v_k, \quad \lambda_k = \frac{6}{h^2} \cdot \frac{2 - 2\cos(k\pi h)}{4 + 2\cos(k\pi h)} - \mu.
\end{aligned}
\tag{2.62}
$$

The solution given by (2.62) is a trajectory of type (2.34), i.e. Theorem 2.14 is applicable.

Choice of initial data

To get comparable results for our experiments in different V_h, $h = \frac{1}{N+1}$, $N = 1, \ldots, \overline{N}$ we choose the initial vector $\mathbf{u}_{h,0}$ corresponding to the initial function $u_{h,0} \in V_h$ as the L^2-projection of a fixed initial function in $V_{\overline{h}}$, $\overline{h} = \frac{1}{\overline{N}+1}$:

$$u_0 = \sum_{i=1}^{\overline{N}} \alpha_i^0 \Lambda_i \in V_{\overline{h}},
\tag{2.63}$$

where we choose $\{\alpha_i^0\}_{1\leq i\leq \overline{N}}$ at random. Observe that this refers to stochastical initial data which should represent the worst-case behavior of solutions. We will come back to another possible choice of initial data at the end of our analysis.

Application of Theorem 2.14 for $\ell = 1$
A Taylor expansion immediately shows that the eigenvalues in (2.62) are of order

$$\lambda_k = \lambda_k(h) = k^2\pi^2 + O(h^2).$$

Hence the stronger assumptions (2.38) of Theorem 2.14 are satisfied.

Let us first take a look at the case $\ell = 1$, i.e. we analyze the behavior of the first POD vector $w_1 \in \mathbb{R}^N$ measured by the angle between w_1 and the first Fourier mode v_1. Using the notation of Theorem 2.14 we get an explicit bound $\gamma = \gamma(T)$ with

$$\gamma = e^{-\delta T}, \quad \delta = \lambda_2 - \lambda_1 \approx 3\pi^2.$$

Since V is orthonormal, we can derive $M = M_N(\bar{u}_{0,N})$ depending on the initial data $\bar{u}_{0,N}$:

$$M = M_N = \frac{|(V\bar{u}_{0,N})_1|}{\sqrt{2}\max_{2\leq k\leq N}|(V\bar{u}_{0,N})_k|}.$$

Therefore the bound (2.36) for the angle is given explicitly by

$$E_N = E_N(T, \bar{u}_{0,N}) = \frac{\gamma(T)}{M_N(\bar{u}_{0,N}) - \gamma(T)}.$$

Figure 2.2 shows the results of the experiments for $\overline{N} = 100$. We plot the angle between the first eigenvector $v_1 \in \mathbb{R}^N$ and the first POD vector as a function of the spatial dimension $1 \leq N \leq 100$ together with the theoretical bound E_N for different values of T all satisfying $T \geq 2/\delta$ as in the theorem. We see that for small N, the canonical angle is increasing but converges very quickly. The theoretical bound holds indeed in all experiments. The numerically derived angle in the experiments is of a smaller magnitude than the theoretical bound—by a factor of about 10-20. Nevertheless the theoretical bound predicts the right order of magnitude of the angle in all cases.

In Figure 2.3 we take a closer look to the beginning of the experiments, i.e. to the interval $1 \leq N \leq 20$. We see that the bound is even sharper when we look at small N. We mention a small numerical artefact in the last plot of Figure 2.2. We see that there is a small interval where the error bound does not hold. This can be explained by roundoff errors if we look at the magnitude of the computed angle.

In Figure 2.4 we compare the results of Figure 2.2 for $T = 0.5$ with an experiment for smoother initial data. In detail the initial data is chosen randomly in the space $V_{\frac{1}{11}}$ instead of $V_{\frac{1}{101}}$ as in (2.63). By that, more or less only the first 10 Fourier modes $v_1, \ldots, v_{10}$ are involved in the L^2-projections for $N \geq 10$.

We see in the plots that, qualitatively, the relation between the canonical angle and the bound given by Theorem 2.14 does not change. Both curves are just getting smoother for $N \geq 10$. This meets our expectations since the higher Fourier modes are negligible in this case.

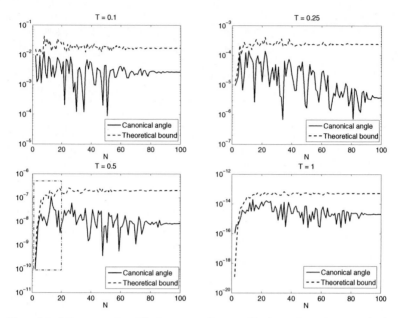

Figure 2.2: A linear reaction-diffusion system discretized by linear finite elements. Angle between first eigen- and POD-vector together with the theoretical bound derived from (2.36). Different values of T.

Application of Theorem 2.14 for $\ell = 2$

We want to extend our experiments to the case of $\ell = 2$ POD modes to illustrate the bounds of Theorem 2.14 also for $\ell > 1$. With analogous calculations as above we derive the bounds for $\gamma = \gamma(T)$ and $M_N = M_N(\mathbf{u}_{0,N})$ also for the case $\ell = 2$. Observe that we have $\delta = \lambda_3 - \lambda_2 \approx 5\pi^2$ with the exact value given by (2.62) and further on

$$\text{gap}_\alpha = |\alpha_1 - \alpha_2| = e^{-\lambda_1 T} - e^{-\lambda_2 T},$$
$$\sigma_2(V_0) = \sigma_2((V\mathbf{u}_0)_1 v_1, (V\mathbf{u}_0)_2 v_2) = \min\{(V\mathbf{u}_0)_1, (V\mathbf{u}_0)_2\},$$
$$\sigma_1(V_1) = \sigma_1(\{(V\mathbf{u}_0)_k v_k\}_{k\geq 3}) = \max\{(V\mathbf{u}_0)_k\}_{k\geq 3}.$$

Again, we can derive the bound $E_N = \frac{\gamma}{M_N - \gamma}$ explicitly with

$$\gamma = e^{-(\lambda_3 - \lambda_2)T}, \quad M_N = \frac{(e^{-\lambda_1 T} - e^{\lambda_2 T})}{4} \frac{\min\{(V\mathbf{u}_0)_1, (V\mathbf{u}_0)_2\}}{\max_{k=3,\dots,N}(V\mathbf{u}_0)_k}.$$

Numerical experiments show that we have to be careful in choosing the integration time T. If we choose T too large (e.g. $T = 1$), the second singular value almost vanishes. By that the problem of finding a second singular vector gets ill-posed such that the second POD mode is chosen almost randomly by the algorithm and does not fit to the snapshots.

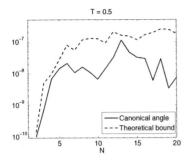

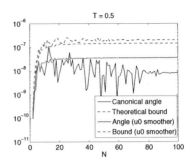

Figure 2.3: Blowup of the first $N = 20$ experiments of the example as above for $T = 0.5$. Plot of the region marked in Figure 2.2.

Figure 2.4: The same system as in figure 2.2 for $T = 0.5$. Comparison of the results for the stochastic initial data as in Figure 2.2 vs. smoother initial data.

In Figure 2.5 we have plotted the experimental data in the same way as in Figure 2.2 for the case $\ell = 2$ and $T \in \{0.1, 0.25, 0.5\}$.

For $T = 0.1$ the bound exists only for some $N \in \{1, \dots, \bar{N}\}$. This is due to the fact that in most of the randomly chosen initial data the condition $M_N \geq \gamma$ is violated by choosing T that small. Therefore in the first plot of Figure 2.5, the theoretical bound is only evaluated in those experiments where $M_N \geq \gamma$ is satisfied.

If we choose a proper T like for example for $T = 0.25$ and $T = 0.5$, also for $\ell = 2$ the bounds of Theorem 2.14 are very suitable. As it is the case for $\ell = 1$, the magnitude of the canonical angles is described quite well by the error bounds. In particular the bounds overestimate the angles only by a factor of about $10 - 20$ as for $\ell = 1$.

2.3.6 Different initial values

The AIM algorithm 1.30 is based on the computation of many short time trajectories. We will see in detail in Chapter 3 how to make use of these short time trajectories for the computation of POD bases. Therefore we end this chapter with a first observation of how the POD method behaves if more than one trajectory is used to define snapshots.

Theorem 2.17. *Consider trajectories $u(t; u_0)$ and $u(t; u_1)$ to different initial values u_0 and u_1 with $\|u_0 - u_1\|_2 \leq \varepsilon$. Assume these trajectories are shadowing each other, i.e.*

$$\|u(t; u_0) - u(t; u_1)\|_2 \leq \varepsilon, \quad t \geq 0.$$

As before, let a collection of snapshots be given by

$$y_j = u(t_j; u_0) = u(jT; u_0), \quad j = 1, \dots, m.$$

Assume a spectral gap $\delta > 0$ for the singular values $\{\sigma_i\}_i$ of $Y = \frac{1}{\sqrt{m}} \operatorname{col}(y_1, \dots, y_m)$:

$$\sigma_\ell - \sigma_{\ell+1} =: \delta > 0.$$

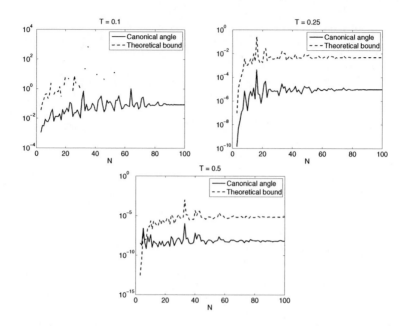

Figure 2.5: Same example as in Figure 2.2. Angle between the first $\ell = 2$ eigen- and POD-vectors together with the theoretical bound derived from (2.36). Different values of T.

Another collection of snapshots is built from both trajectories:

$$x_j = y_j = u(jT; u_0), \ x_{m+j} = z_j = u(jT; u_1), \quad j = 1, \dots, m.$$

Then as an estimate for the angle between the POD space $\mathcal{W}^Y$ of rank ℓ for the snapshots $\{y_j\}_{j=1}^m$ and the POD space $\mathcal{W}^X$ referring to the snapshots $\{x_j\}_{j=1}^{2m}$, we get:

$$\| \sin(\angle(\mathcal{W}^Y, \mathcal{W}^X)) \|_2 \leq \frac{\varepsilon}{\sqrt{2}\delta - \varepsilon} \leq \sqrt{2}\frac{\varepsilon}{\delta}$$

where the second estimate holds for $\varepsilon < \frac{\delta}{\sqrt{2}}$.

Proof. Let the singular value decomposition of $Y = \frac{1}{\sqrt{m}} \mathrm{col}(y_1, \dots, y_m)$ be given by

$$W^T Y V = D := \left(\mathrm{diag}(\sigma_1, \dots, \sigma_N) \quad 0 \right)$$

with $\sigma_1 \geq \dots \geq \sigma_N \geq 0$. Observe that for the matrix

$$Y_2 = \frac{1}{\sqrt{2m}} \mathrm{col}(y_1, \dots, y_m, y_1, \dots, y_m) = \frac{1}{\sqrt{2}} \left(Y \quad Y \right)$$

we get the same singular values as for Y. A short calculation shows that the matrix

$$V_2 := \frac{1}{\sqrt{2}} \begin{pmatrix} V & V \\ V & -V \end{pmatrix} \in \mathbb{R}^{2m,2m}$$

is orthogonal. With $W \in \mathbb{R}^{N,N}, D \in \mathbb{R}^{N,m}$ as above we get

$$W^T Y_2 V_2 = \frac{1}{2} W^T \begin{pmatrix} Y & Y \end{pmatrix} \begin{pmatrix} V & V \\ V & -V \end{pmatrix}$$

$$= \frac{1}{2} \begin{pmatrix} W^T Y & W^T Y \end{pmatrix} \begin{pmatrix} V & V \\ V & -V \end{pmatrix} = \frac{1}{2} \begin{pmatrix} 2W^T Y V & 0 \end{pmatrix} = \begin{pmatrix} D & 0 \end{pmatrix}.$$

If we define $X = \frac{1}{\sqrt{2m}} \operatorname{col}(y_1, \ldots, y_m, z_1, \ldots, z_m)$, we get the norm-wise estimate

$$\|X - Y_2\|_2 \le \|X - Y_2\|_F = \frac{1}{\sqrt{2m}} \left(\sum_{j=1}^{m} \|y_j - z_j\|_2^2 \right)^{1/2} \le \frac{\varepsilon}{\sqrt{2}}.$$

In this way we can compare the POD space $\mathcal{W}^Y$ of rank $\ell \le m$ associated to the snapshots $\{y_i\}_i$ along one trajectory with the POD space $\mathcal{W}^X$ associated to the snapshots $\{y_i, z_i\}_i$ along two trajectories and get with Corollary A.13:

$$\|\sin(\angle(\mathcal{W}^Y, \mathcal{W}^X))\|_2 \le \frac{\|X - Y_2\|_2}{\delta - \|X - Y_2\|_2} \le \frac{\varepsilon}{\sqrt{2}\delta - \varepsilon}.$$

In the case $\varepsilon < \frac{\delta}{\sqrt{2}}$ we can simplify the denominator to get

$$\|\sin(\angle(\mathcal{W}^Y, \mathcal{W}^X))\|_2 \le \sqrt{2}\frac{\varepsilon}{\delta}.$$

$\square$

Remark. *Theorem 2.17 shows that two shadowing trajectories essentially do not generate more information than a single trajectory in the sense that the resulting POD modes are close to each other.*

In our concluding example we try to broaden the results of Theorem 2.17 by numerical means. The setting of the following computation is not covered any more by the assumptions of the theorem. Yet we mention it here since it gives a nice prospect for the POD algorithms where POD modes are computed from a whole bundle of short time trajectories. In the computations for Figure 2.6, we derive the first POD vector for the discretized linear parabolic system (2.62) with $T = 0.5$ in different ways. We build three different sets of snapshots collected in

$$Y^i = \frac{1}{\sqrt{6m}} \operatorname{col}(y_1^i, \ldots y_{6m}^i), \quad i = 1, 2, 3.$$

We consider trajectories $u^i(t) = u(t, \mathbf{u}_0^i)$ starting in 3 different initial points $\mathbf{u}_0^i$, $i = 1, 2, 3$, with randomly chosen coefficients as above. The first collection of snapshots is built from one trajectory of length $6mT$ as above:

$$y_j^1 = u^1(t_j), \quad j = 1, \ldots, 6m.$$

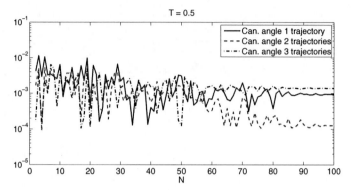

Figure 2.6: The same linear reaction-diffusion system discretized by linear finite elements as in Figure 2.2. Angle between the first eigen- and POD-vector for snapshots taken from one trajectory, two trajectories and three trajectories in the case of $T = 0.5$.

The second collection is built from two trajectories of half length $3mT$:

$$y_j^2 = u^1(t_j), \quad y_{3m+j}^j = u^2(t_j), \quad j = 1, \ldots, 3m.$$

Finally, the last collection of snapshots is built from three trajectories of length $2mT$:

$$y_j^3 = u^1(t_j), \quad y_{2m+j}^3 = u^2(t_j), \quad y_{4m+j}^3 = u^3(t_j), \quad j = 1, \ldots, 2m.$$

The resulting angles between the first POD mode w_1^i, $i = 1, 2, 3$, and the first Fourier mode in a typical realization of the experiment is plotted in Figure 2.6. We should mention that the computed angles depend strongly on the choice of the initial data u_0^i. We have shown here a typical curve shape.

Obviously, the first eigendirection is approximated better by two trajectories than by one, at least for large N. Nevertheless for three trajectories in most realizations the error gets worse again, as it is the case in the example shown in Figure 2.6. It seems that the length of the involved trajectories is of greater advantage in this case than the convergence from more than two initial points in view of the approximation of the eigendirections. We just predict here that in cases of more complicated dynamics, it will be more benefitting to use many trajectories, see Section 6.2 for more details. A deeper theoretical analysis of these phenomena would go beyond the scope of this thesis and is part of a future work.

Chapter 3

Set-valued algorithms in reduced space

In Section 1.4 we have introduced the AIM algorithm. It is a subdivision algorithm computing discrete measures as approximations to an SRB measure defined in Section 1.2. This technique is a powerful tool to approximate invariant measures with high accuracy for low-dimensional discrete dynamical systems—think of models defined by low-dimensional maps or of discretizations of low-dimensional ordinary differential equations.

In this thesis we want to use this technique for the computation of invariant measures in parabolic problems. Since discretizations, as e.g. by the finite element method described in Chapter 2, result in high-dimensional systems, the original version of the algorithm suffers from the so-called 'curse of dimension':

Even in cases where the support of the invariant measure is known to be low-dimensional, e.g. in the Chafee-Infante problem, see Chapter 6, the adaptive algorithm 1.30 produces an exponentially growing number of boxes in the first recursion steps. Here, the adaptive refinement of boxes in step (1.17) does not compensate this exponentially growth. Hence, after one bisection in every spatial component $1 \leq k \leq N$ the support of the discrete measures consists of nearly 2^N boxes and the algorithm breaks down for large N after only a few recursions.

Therefore, it is a promising approach to combine the subdivision technique of Algorithm 1.30 with model reduction. One approach for model reduction is given by the POD method which we described and analyzed in Chapter 2. We introduce two algorithms arising out of these considerations in the following.

3.1 PODAIM: The Invariant Measure Algorithm in reduced space

As before, we consider a discrete dynamical system defined by a sufficiently smooth function $F : X \to X$, $X \subset \mathbb{R}^N$ compact,

$$u_{i+1} = F(u), \qquad i = 0, 1, \ldots \tag{3.1}$$

where $N \in \mathbb{N}$ is supposed to be large. Because of the 'curse of dimension' described above, the number of support boxes in Algorithm 1.30 exceeds a manageable amount after just a few recursion steps of the algorithm. Therefore it is an obvious approach to do a model reduction before applying subdivision techniques. We follow this ansatz in algorithm 3.1 by use of the POD method. We will see later on in numerical examples that this yields a feasible ansatz for some parabolic problems such as the Chafee-Infante problem.

The algorithm computes discrete measures in the POD space. At the end of the algorithm we show how the information provided by the AIM algorithm in reduced space can be used to approximate invariant measures in the original system given by (3.1).

Algorithm 3.1. The PODAIM algorithm

- **Snapshots:** As in algorithm 1.30, let $B_0^{(N)} \subset X$ be a positive invariant box. For randomly chosen points $u_{0,1}, \ldots, u_{0,m} \in B_0^{(N)}$, $m \geq N$, compute short trajectories of a time $T \in \mathbb{N}$ to get a collection of snapshots

$$y_i = F^T(u_{0,i}), \ i = 1 \ldots, m. \tag{3.2}$$

- **POD modes:**

    - Compute the singular value decomposition of $Y = \frac{1}{\sqrt{m}} \operatorname{col}(y_1, \ldots, y_m) \in \mathbb{R}^{N,m}$:

$$W^T Y U = \begin{pmatrix} \operatorname{diag}(\sigma_1, \ldots, \sigma_N) & 0 \\ 0 & 0 \end{pmatrix}$$

    with $\sigma_1 \geq \ldots \geq \sigma_N \geq 0$ and $W \in \mathbb{R}^{N,N}, V \in \mathbb{R}^{m,m}$ orthogonal.
    - Choose $\ell \ll N$ and split W into $W_1 \in \mathbb{R}^{N,\ell}$, $W_2 \in \mathbb{R}^{N,N-\ell}$ such that $W = (W_1, W_2)$.

- **AIM algorithm in POD space:** Apply the AIM algorithm to the system

$$\alpha_{i+1} = F_{\mathrm{red}}(\alpha_i), \qquad i = 0, 1, \ldots \tag{3.3}$$

where $F_{\mathrm{red}} : \mathbb{R}^\ell \to \mathbb{R}^\ell$ is defined by

$$F_{\mathrm{red}}(\alpha) = W_1^T F(W_1 \alpha). \tag{3.4}$$

For this, choose a proper starting box $B_0^{(\ell)}$ in $\mathbb{R}^\ell$ that is positive invariant for F_{red}. In the k-th recursion step of the AIM algorithm, obtain $\mu_k^{(\mathrm{red})} : \mathcal{B}(\mathbb{R}^\ell) \to [0,1]$ defined by

$$\mu_k^{(\mathrm{red})}(A) = \sum_{i=1}^K u_i \frac{\lambda_\ell(A \cap B_i)}{\lambda_\ell(B_i)}, \quad A \in \mathcal{B}(\mathbb{R}^\ell) \tag{3.5}$$

where $\mathcal{B}_k = \{B_i\}_{i=1}^K$ is the current box collection in the reduced space.

- **Extension to original space:** At the end, a support box $B = B(c,r) \subset \mathbb{R}^\ell$ corresponds to the set $W_1(B) \subset \mathbb{R}^N$ in the hyperplane defined by W_1:

$$W_1(B) := \{z = W_1 x \in \mathbb{R}^N : x \in B\}.$$

These embedded boxes form the support of the *extended measure* $\mu_k^{(\mathrm{pod})} : \mathcal{B}(\mathbb{R}^N) \to [0,1]$ defined by

$$\mu_k^{(\mathrm{pod})}(A) = \sum_{i=1}^K u_i \frac{\lambda_\ell(W_1^{-1}(A) \cap B_i)}{\lambda_\ell(B_i)} \tag{3.6}$$

where $W_1^{-1}(A) := \{x \in \mathbb{R}^\ell : W_1 x \in A\}$ is the preimage of A under the linear transformation defined by $W_1 \in \mathbb{R}^{\ell,N}$.

Remark. *There is some freedom in choosing the number ℓ of POD modes in the algorithm. It is reasonable to search for an index $\ell > 0$ with large spectral gap $\sigma_\ell - \sigma_{\ell+1}$. A suitable approach is given by choosing the smallest index with*

$$\frac{\sigma_{\ell+1}}{\sigma_\ell} \leq tol \tag{3.7}$$

where $tol \geq 0$ is a given tolerance.

Since we operate in a subspace of the state space it is obvious that we cannot expect absolutely continuous measures with respect to the Lebesgue measure λ_N. Indeed, the next theorem shows that supports of the discrete measures are thin in $\mathbb{R}^N$.

Proposition 3.2. *With the notations of algorithm 3.1 let the extended measure $\mu_k^{(\text{pod})}$: $\mathcal{B}(\mathbb{R}^N) \to [0,1]$ be defined by (3.6).*

1. *Define the Lebesgue measure λ_ℓ^W on the ℓ-dimensional subspace $\text{span}\{w_1, \ldots, w_\ell\} \subset \mathbb{R}^N$ by*

$$\lambda_\ell^W(A) = \lambda_\ell(W_1^{-1}(A)), \quad A \in \mathcal{B}(\mathbb{R}^N)$$

Then $\mu_k^{(\text{pod})}$ can be written as

$$\mu_k^{(\text{pod})}(A) = \sum_{i=1}^K u_i \frac{\lambda_\ell^W(A \cap W_1(B_i))}{\lambda_\ell^W(W_1(B_i))}, \quad A \in \mathcal{B}(\mathbb{R}^N) \tag{3.8}$$

2. *With the Dirac measure $\delta_0^{N-\ell} : \mathcal{B}(\mathbb{R}^{N-\ell}) \to [0,1]$ at 0, defined in (A.7), it holds*

$$\mu_k^{(\text{pod})} = \left(\mu_k^{(\text{red})} \times \delta_0^{N-\ell}\right) \circ W^T \tag{3.9}$$

where $\mu \times \nu$ is the product measure of μ and ν defined in Definition A.22 and $\mu \circ W$ is the linear transformation of the distribution μ given by (A.20).

Proof.   1. The representation (3.8) is just a reformulation of (3.6) by observing that

$$W_1^{-1}(W_1(B_i)) = B_i \quad \text{and}$$
$$W_1^{-1}(A \cap W_1(B_i)) = W_1^{-1}(A) \cap B_i.$$

2. Let $A \in \mathcal{B}(\mathbb{R}^N)$ be given.

By (A.20) and Fubini's Theorem A.28 for $\nu = \left(\mu_k^{(\text{red})} \times \delta_0^{N-\ell}\right) \circ W^T$, we get

$$
\begin{aligned}
\nu(A) &= \int \mathbb{1}_A \, d\nu \overset{(A.20)}{=} |\det W^T|^{-1} \int \mathbb{1}_A(Wx) \, d(\mu_k^{(\text{red})} \times \delta^{N-\ell})(x) \\
&= \int \mathbb{1}_{W^T(A)}(x) \, d(\mu_k^{(\text{red})} \times \delta^{N-\ell})(x) = (\mu_k^{(\text{red})} \times \delta_0^{N-\ell})(W^T(A)) \\
&\overset{(\text{Thm. A.28})}{=} \int \delta_0^{N-\ell}(\{y \in \mathbb{R}^{N-\ell} : (x,y) \in W^T(A)\}) \, d\mu_k^{(\text{red})} \\
&= \int \mathbb{1}_{W_1^{-1}(A)} \, d\mu_k^{(\text{red})} = \mu_k^{(\text{red})}(W_1^{-1}(A)) \\
&\overset{(3.5)}{=} \sum_{i=1}^K u_i \frac{\lambda_\ell(W_1^{-1}(A) \cap B_i)}{\lambda_\ell(B_i)}.
\end{aligned}
$$

$\square$

3.2 Adaptive POD algorithm

Looking closer at the original subdivision algorithm, it is an obvious idea to compute the POD modes adaptively during the recursion steps of the AIM algorithm 1.30. Since the algorithm computes short time trajectories to build the transition matrix as in (1.19), the same heuristic can be used to derive an adapted POD basis during the algorithm.

The adaptive algorithm PODADAPT works as follows. Instead of computing discrete measures in a fixed state space, as well the POD subspace will be changed dynamically. Therefore we manage not only a box collection and a corresponding discrete measure $(\mathcal{B}_k, \mu_k)$ throughout the algorithm as in Algorithm 1.30 but also the POD space given by (ℓ_k, W_k). Here, $\ell_k \in \mathbb{N}$ denotes the dimension and $W_k \in \mathbb{R}^{N, \ell_k}$ describes the basis of the POD space. We start with the original state space, or in other words the canonical basis of $\mathbb{R}^N$ as the first POD basis. After some recursion steps we compute a new POD basis. Now the system is transformed into the new state space and a new box collection of comparable complexity is built in the new state space. Then the algorithm continues to work in the new state space. This leads to the following algorithm.

Algorithm 3.3. The PODADAPT algorithm

- **Initialization:** As in the AIM algorithm 1.30, start with a positive invariant box $B_0^{(N)} \subset X$, the box collection $\mathcal{B}_0 = \{B_0^{(N)}\}$ and the discrete measure $\mu_0^{(\text{red})}$ with constant density on $B_0^{(N)} = \operatorname{supp} \mu_0^{(\text{red})}$. The initial POD space is given by $\mathbb{R}^N$, i.e. $\ell_0 = N$ and $W_0 = I_N$.

- **Recursion step k:** Let $(\mathcal{B}_{k-1}, \mu_{k-1}^{(\text{red})}, \ell_{k-1}, W_{k-1})$ be given.

  1. *Subdivision step:* In the POD space given by (ℓ_{k-1}, W_{k-1}), perform an AIM recursion step as in Algorithm 3.1 to obtain a refined box collection $\widetilde{\mathcal{B}}_k$ of size $K = \#\widetilde{\mathcal{B}}_k$ with a discrete measure $\widetilde{\mu}_k$ corresponding to $\widetilde{\mathcal{B}}_k$.

  2. *POD transformation:* After a given number P of recursion steps compute a new POD basis by the method given below. Alternatively, set

$$(\mathcal{B}_k, \mu_k^{(\text{red})}, \ell_k, W_k) = (\widetilde{\mathcal{B}}_k, \widetilde{\mu}_k, \ell_{k-1}, W_{k-1}),$$

  and continue with the next recursion.

  - **Snapshots:** In each box $B_j \in \widetilde{\mathcal{B}}_k$ take randomly chosen points $u_{j,\alpha} \in B_j$, $\alpha = 1, \ldots, m_j$, where m_j is determined by the discrete measure $\widetilde{\mu}_k$:

$$m_j = \left\lceil \frac{\widetilde{\mu}_k(B_j)}{\max\limits_{i=1,\ldots,K} \widetilde{\mu}_k(B_i)} \cdot m \right\rceil. \qquad (3.10)$$

  Here $m \in \mathbb{N}$ is fixed and $\lceil x \rceil$ is the smallest natural number greater than or equal to x. Take trajectories of time $T \in \mathbb{N}$ as in (3.2) to get a collection of snapshots

$$y_{j,\alpha} = F^T(u_{j,\alpha}), \quad \alpha = 1, \ldots, m_j, \ j = 1, \ldots, K. \qquad (3.11)$$

- **POD modes:** Calculate a new POD space (ℓ_k, W_k), $W_k \in \mathbb{R}^{N,\ell_k}$ by a singular value decomposition of

$$Y = \frac{1}{\sqrt{\sum_{j=1}^{K} m_j}} \operatorname{col}(\{y_{j,\alpha} : \alpha = 1, \ldots, m_j, \ j = 1, \ldots, K\}).$$

Choose ℓ_k reasonably as in Algorithm 3.1.

- **Transformation:** Transfer the system to the new POD space using the following steps

  (a) Determine the box $B = B(c,r) \in \widetilde{\mathcal{B}}_k$ with the smallest diameter of the current collection and set $r_{\min} := \min\{r_i : i = 1, \ldots, \ell_{k-1}\}$.

  (b) Create a finite box collection $\widetilde{\mathcal{B}}_k^W = (\widetilde{B}_i^W)_i$ as a covering of $W_k^T B_0^{(N)}$ where all boxes $\widetilde{B}_i^W = B(c_i, r)$ have the same radius:

  $$r = r_{\min}\mathbb{1} \quad \text{with } \mathbb{1} = (1, \ldots, 1)^T.$$

  (c) Derive a new discrete measure $\widetilde{\mu}_k : \mathcal{B}(\mathbb{R}^{\ell_k}) \to [0, 1]$ from the distribution of the embedded snapshots: For $B \in \widetilde{\mathcal{B}}_k^W$ define

  $$\widetilde{\mu}_k(B) = \Big(\sum_{j=1}^{K} m_j\Big)^{-1} \#\{W_k y_{j,\alpha} \in B : \alpha = 1, \ldots, m_j, \ j = 1, \ldots, K\}.$$

  (d) As in the AIM algorithm, remove boxes with null measure from the collection to obtain

  $$\mathcal{B}_k = \{B \in \widetilde{\mathcal{B}}_k : \widetilde{\mu}_k > 0\}$$

  and adjust the measure

  $$\mu_k^{(\mathrm{red})} = (\widetilde{\mu}_k)_{|\mathcal{B}_k}.$$

Remark. *In the algorithm there is some freedom in the choice of parameters. As before, the number ℓ_k of POD modes defining the new space has to be chosen adequately. As in Algorithm 3.1 a proper ansatz is given by the criterion (3.7).*

Secondly for the snapshot collection, the number $m \in \mathbb{N}$ of test points has to be chosen in (3.10) as well as the integration time $T \in \mathbb{N}$ for the trajectories in (3.11). It is reasonable to choose m a bit larger than in Algorithm 3.1 in order to compensate that $\operatorname{supp} \widetilde{\mu}_k \subset \mathbb{R}^{\ell_{k-1}}$ is in generally small in the sense that it is a Lebesgue null set in the original state space $\mathbb{R}^N$ for $\ell_{k-1} < N$.

Later on in Chapter 6 we will analyze the behavior of our algorithm in more detail. We already state here that it turns out to be a hard task to set up the PODADAPT algorithm properly such that no dynamical behavior is lost during the adaptation of the POD space. Therefore in most cases it is appropriate to use the less sophisticated ansatz of the PODAIM algorithm which produces promising results by working in a fixed POD space.

3.3 Analysis of the algorithms: An outlook

Having introduced algorithms that compute discrete measures in high-dimensional state spaces $\mathbb{R}^N$, $N \gg 1$, the question arises of how we can display the information represented by these measures. To complicate things further, the supports of the discrete measures computed in the algorithms 3.1 and 3.3 are in general subsets of a hyperplane that varies for different runs of the program or even between different recursion steps. We recall that these discretized dynamical systems usually come from a full discretization of a parabolic problem with finite elements used for the spatial discretization. Hence it is an obvious approach to visualize the measures in the state space of the partial differential equation. In Section 6.1 we will follow this approach by introducing a histogram-like representation for discrete measures corresponding to scalar partial differential equations.

For the analysis of the above algorithms it is important to develop a concept of comparing discrete measures with special geometry as above. It turns out that the common definition of the weak metric d_w given by (A.15) is not suitable in our case. Numerical discretizations of the weak metric do not work in our context because the definition disregards the special geometry of the measures. We will show in Section 4.1 why the Prohorov metric which is equivalent to the weak metric, is a suitable distance of measures that can be computed numerically very well.

Although a convergence theory for the reduced-space algorithm is well out of reach, in Chapter 5 we will present some error estimates concerning the discrete measures derived by the algorithms. Finally in Chapter 6, we will see in various numerical experiments, that the algorithms are particularly suitable for the computation of invariant measures in high-dimensional systems where the support of an invariant measure is known to be part of a low-dimensional manifold.

Chapter 4

Comparison of discrete measures

In this chapter, we derive a proper distance notion for the discrete measures arising in the POD-based algorithms in the previous chapter. We will see that the Prohorov metric is the most promising choice since it takes into account the geometry of the discrete measures. We will compare this distance notion with the usual weak metric for probability measures on a compact metric space and with the negative Sobolev norm. At the end, we present contraction results for the Perron-Frobenius matrix in different norms and metrices and in particular in the Prohorov metric.

4.1 Distance notions for discrete measures—the Prohorov metric

In the preceding chapter, we have introduced algorithms for the approximation of invariant measures in discrete dynamical systems of type (3.1) with space dimension $N \gg 1$. In the algorithms 3.1 and 3.3, the geometry of the approximating discrete measures is described by (3.5). Observe that the support is given by a low-dimensional box collection $\mathcal{B} = \{B_1, \ldots, B_K\}$ embedded in an ℓ-dimensional subspace span W of the state space $\mathbb{R}^N$.

For the further analysis of our algorithms it is necessary to develop a proper concept of comparing such measures in different POD spaces given by $W \in \mathbb{R}^{N,\ell}$. Also a discrete measure derived from the AIM algorithm 1.30 can be interpreted as a discrete measure in a special POD space given by the matrix $W = I_N$. Hence we can in particular compare the results of POD-based algorithms with the results of the AIM algorithm by this concept.

There are different approaches for computing such distances between measures. We introduce three of them in the following. As denoted above, the Prohorov metric is the proper distance notion in our case. (see Sections 4.1.2, 4.1.3). But also the use of negative Sobolev norms is conceivable from a theoretical point of view, as we will see in Section 4.1.4. However, we show in Section 4.1.5 that the numerical realization is much too expensive if the state space dimension N is large.

We begin our investigations with a common approach for a distance notion on the space of measures, the *weak metric*. We explain why we reject this approach for a distance notion of discrete measures.

4.1.1 The weak metric

It is well-known that the space $\mathcal{M}^1(B_0)$ of probability measures on a compact set $B_0 \in \mathbb{R}^N$ is a compact metric space with weak metric d_w defined by

$$d_w(\mu, \nu) = \sum_{i=0}^{\infty} 2^{-i} \left| \int g_i \, d\mu - \int g_i \, d\nu \right|, \quad \mu, \nu \in \mathcal{M}^1(B_0), \tag{4.1}$$

where $(g_i)_i$ is a dense sequence in $C(B_0)$ (cf. Theorem A.38).

If we use this metric in our numerical experiments we have to find a proper discretization of (4.1) for the case where μ and ν have support on box collections $\{A_j\}_j$, $\{B_k\}_k$ in the POD spaces defined by $W_1 \in \mathbb{R}^{N,\ell_1}$ and W_2^{N,ℓ_2}. It is a natural approach to use test functions according to the geometry of our discrete measures, i.e. the characteristic functions

$$\left\{ \mathbb{1}_{W_1(A_1)}, \ldots, \mathbb{1}_{W_1(A_J)}, \mathbb{1}_{W_2(B_1)}, \ldots, \mathbb{1}_{W_2(B_K)} \right\}.$$

However, this choice is very sensitive to small perturbations. In particular, the distance in such a discrete version of the weak metric equals 1 as soon as span $W_1 \neq$ span W_2. Hence, the weak metric as defined in (4.1) is not stable under numerical discretizations and so we avoid to use this definition directly.

4.1.2 The concept of blowing up boxes

A promising ansatz is given by the Prohorov metric mainly used in probability theory to develop convergence concepts, cf. [Dud02], [Bil99]. It is also used in the numerical analysis of random dynamical systems, cf. [IK03], [DKP95]. The abstract definition on a metric space is as follows:

Definition 4.1. Let (S, d) be a metric space. Denote by A^ε, $\varepsilon > 0$, the ε-hull of a subset $A \subset S$

$$A^\varepsilon := \{y \in S : d(x, y) < \varepsilon \text{ for some } x \in A\}, \tag{4.2}$$

cf. Definition 1.5. Then the *Prohorov metric* on the set $\mathcal{M}^1(S)$ of probability measures on S is defined by

$$p : \mathcal{M}^1(S) \times \mathcal{M}^1(S) \to \mathbb{R}_+,$$
$$p(\mu, \nu) = \inf\{\varepsilon > 0 : \mu(A) \leq \nu(A^\varepsilon) + \varepsilon \text{ for all } A \in \mathcal{B}(S)\}. \tag{4.3}$$

Remark. *Let us collect a few properties of the Prohorov metric.*

- *By topological arguments, it is easy to show that p is indeed a metric on $\mathcal{M}^1(S)$. In particular, it does not make a difference if we swap the order of the measures in (4.3):*

$$p(\mu, \nu) = \inf\{\varepsilon > 0 : \mu(A) \leq \nu(A^\varepsilon) + \varepsilon \text{ and } \nu(A) \leq \mu(A^\varepsilon) + \varepsilon \text{ for all } A \in \mathcal{B}(S)\}.$$

- *The metric combines a distance in the state space by blowing up boxes by the factor ε with a distance in the measures by adding ε.*

- *In probability theory, the main application of this distance is given by the classification of convergence concepts for probability measures and random variables.*

- *The blow-up concept of the Prohorov distance perfectly fits into our setting: Contrary to the weak metric, small perturbations of the POD space correspond to small ε-hulls of boxes. Thus, p is well-suited for numerical realizations. We will define a proper discretization in the following section.*

- *The Prohorov metric is equivalent to the weak metric defined in* (4.1) *in our context:*

Theorem 4.2. *Suppose S is separable and complete. Then weak convergence is equivalent to p-convergence, $\mathcal{M}^1(S)$ is separable and complete, and $A \subset \mathcal{B}(S)$ is relatively compact if and only if the p-closure of A is p-compact.*

Proof. see [Bil99].   □

4.1.3   Implementation of the Prohorov metric

As proposed for the weak metric, we choose the support boxes of the involved measures as test sets for the numerical realization of the Prohorov metric. We recall the definition of discrete measures $\mu_k^{(\text{pod})}, \nu_k^{(\text{pod})} : \mathcal{B}(\mathbb{R}^N) \rightarrow [0,1]$: Given the support box collections $\{A_j\}_{j=1}^{K_\mu} \subset \mathbb{R}^{\ell_1}$ and $\mathcal{B} = \{B_i\}_{i=1}^{K_\nu} \subset \mathbb{R}^{\ell_2}$ embedded into $\mathbb{R}^N$ by $W_1 \in \mathbb{R}^{N,\ell_1}, W_2 \in \mathbb{R}^{N,\ell_2}$ the discrete measures in subdivision step k are given by

$$
\begin{aligned}
\mu(A) = \mu_k^{(\text{pod})}(A) &= \sum_{j=1}^{K_\mu} u_j \frac{\lambda_{\ell_1}(A_j \cap W_1^{-1}(A))}{\lambda_{\ell_1}(A_j)}, \\
\nu(A) = \nu_k^{(\text{pod})}(A) &= \sum_{i=1}^{K_\nu} v_i \frac{\lambda_{\ell_2}(B_i \cap W_2^{-1}(A))}{\lambda_{\ell_2}(B_i)}
\end{aligned}
\tag{4.4}
$$

for every Borel set A. While almost all of the following computations can be done analytically we need two approximation steps for the discretization of the Prohorov distance $p(\mu, \nu)$:

1. Replace the test on all measurable sets $A \in \mathcal{B}(\mathbb{R}^N)$ by the test on all support boxes $\{A_j\}_j, \{B_i\}_i$.

2. Approximate the measure of the blow-up of an embedded support box $W_1(A)^\varepsilon$ by a Monte-Carlo approach.

By the first approximation step we get

$$
p(\mu, \nu) \approx \tilde{p}(\mu, \nu) := \max\{p_1(\mu, \nu), p_2(\mu, \nu)\}
\tag{4.5}
$$

with $p_1(\mu, \nu) = \inf\{\varepsilon > 0 : \mu(W_1(A_j)) \le \nu(W_1(A_j)^\varepsilon) + \varepsilon, j = 1, \dots, K_\mu\}$

and $p_2(\mu, \nu) = \inf\{\varepsilon > 0 : \nu(W_2(B_i)) \le \mu(W_2(B_i)^\varepsilon) + \varepsilon, i = 1, \dots, K_\nu\}.$

By definition, we have $\mu(W_1(A_j)) = u_j$ and $\nu(W_2(B_i)) = v_i$, cf. (4.4). Hence we get

$$
p_1(\mu, \nu) = \inf\{\varepsilon > 0 : \max_{j=1,\dots,K_\mu} (u_j - \nu(W_1(A_j)^\varepsilon) - \varepsilon) \le 0\}
$$

$$
\text{and } p_2(\mu, \nu) = \inf\{\varepsilon > 0 : \max_{i=1,\dots,K_\nu} (v_i - \mu(W_2(B_i)^\varepsilon) - \varepsilon) \le 0\}.
$$

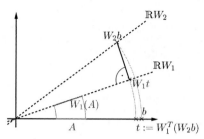

Figure 4.1: Illustration of the Euclidean distance computation needed for the approximation of the Prohorov distance. Here $N = 1$, $\ell_1 = \ell_2 = 1$.

Secondly, we use a Monte-Carlo approach for the approximation of

$$\nu(W_1(A_j)^\varepsilon) = \sum_{i=1}^{K_\nu} v_i \frac{\lambda_{\ell_2}(B_i \cap W_2^{-1}(W_1(A_j)^\varepsilon))}{\lambda_{\ell_2}(B_i)}$$

by choosing test points $b_{i,1}, \ldots, b_{i,P}$ in B_i, $1 \le i \le K_\nu$, and approximating (cf. (1.19))

$$
\begin{aligned}
\frac{\lambda_{\ell_2}(B_i \cap W_2^{-1}(W_1(A_j)^\varepsilon))}{\lambda_{\ell_2}(B_i)} &= \frac{1}{\lambda_{\ell_2}(B_i)} \int_{B_i} \mathbb{1}_{W_2^{-1}(W_1(A_j)^\varepsilon)}(x)\, dx \\
&= \frac{1}{\lambda_{\ell_2}(B_i)} \int_{B_i} \mathbb{1}_{(W_1(A_j)^\varepsilon)}(W_2 x)\, dx \\
&\approx \frac{1}{P} \sum_{p=1}^{P} \mathbb{1}_{W_1(A_j)^\varepsilon}(W_2 b_{i,p}) \\
&= \frac{1}{P}\#\{b_{i,p} : W_2 b_{i,p} \in W_1(A_j)^\varepsilon, p = 1, \ldots, P\} \\
&= \frac{1}{P}\#\{b_{i,p} : d(W_2 b_{i,p}, W_1(A_j)) < \varepsilon, p = 1, \ldots, P\}.
\end{aligned}
$$

Then measuring a blow-up box yields:

$$\nu(W_1(A_j)^\varepsilon) \approx \mathbf{v}_j(\varepsilon) := \sum_{i=1}^{K_\nu} \frac{v_i}{P}\#\{b_{i,p} : d(W_2 b_{i,p}, W_1(A_j)) < \varepsilon, p = 1, \ldots, P\}. \qquad (4.6)$$

Accordingly, we get for μ:

$$\mu(W_2(B_i)^\varepsilon) \approx \mathbf{u}_i(\varepsilon) := \sum_{j=1}^{K_\mu} \frac{u_j}{P}\#\{a_{j,p} : d(W_1 a_{j,p}, W_2(B_i)) < \varepsilon, p = 1, \ldots, P\}.$$

The Euclidean distance $d(W_2 b, W_1(A))$ with W_1, W_2 as above and an ℓ_1-dimensional box $A = B(c,r)$ with center $c \in \mathbb{R}^{\ell_1}$ and radius $r \in \mathbb{R}^{\ell_1}$ can be computed analytically. Figure 4.1 illustrates the geometric approach.

With $t := W_1^T(W_2(b)) \in \mathbb{R}^{\ell_1}$ we can use Pythagoras, since $W_2 b - W_1 t$ is orthogonal to span W_1. We get

$$d(W_2 b, W_1(A))^2 = \underbrace{d(W_1 t, W_1(A))^2}_{=d(t,A)^2} + \|W_2 b - W_1 t\|_2^2.$$

Again using orthogonality, the second term transforms to

$$\|W_2 b - W_1 t\|^2 = \|W_2 b\|^2 - \|W_1 t\|^2 = \|b\|^2 - \|t\|^2$$

while the geometry of $A = B(c,r)$ yields $d(t,A) = \|d\|_2$ for $d = (d_1, \ldots, d_{\ell_1})^T$, $d_i = \max(0, |t_i - c_i| - r_i)$. Together we obtain

$$d(W_2 b, W_1(A)) = \sqrt{\|d\|^2 + \|b\|^2 - \|t\|^2}. \tag{4.7}$$

Observe that only the lengths of low-dimensional vectors $b \in \mathbb{R}^{\ell_2}, d, t \in \mathbb{R}^{\ell_1}$ are used for the computation.

Analogously, we compute $d(W_1 a, W_2(B))$ to derive $\mathbf{u}_i(\varepsilon)$. We then approximate $p_i(\mu, \nu)$ by the first root of the scalar functions $f_i : \mathbb{R}_+ \to \mathbb{R}$, $i = 1, 2$ defined by

$$f_1(\varepsilon) = \max_{j=1,\ldots,K_\mu} (u_j - \mathbf{v}_j(\varepsilon)) - \varepsilon, \quad f_2(\varepsilon) = \max_{i=1,\ldots,K_\nu} (v_i - \mathbf{u}_i(\varepsilon)) - \varepsilon. \tag{4.8}$$

Note that f_1, f_2 are not continuous, since $\mathbf{u}_i(\varepsilon)$, $\mathbf{v}_j(\varepsilon)$ are monotone step functions. However, both functions are strictly antitone, since $\varepsilon > 0$ is subtracted after building the maxima. Thus, good convergence results for root finders can be expected.

Having made these considerations, we can formulate the algorithm for the numerical computation of the Prohorov metric.

Algorithm 4.3. Given two discrete measures $\mu_k^{(\text{pod})}, \nu_k^{(\text{pod})} : \mathcal{B}(\mathbb{R}^N) \to [0,1]$ as in (4.4) approximate the Prohorov distance as follows.

1. Randomly, choose P test points $a_{j,p} \in A_j$, $b_{i,p} \in B_i$, $p = 1, \ldots, P$ in every support box A_j of $\mu_k^{(\text{pod})}$ and B_i of $\nu_k^{(\text{pod})}$.

2. Analytically compute the distance $d(W_1 a, W_2(B))$ for each test point $a \in \text{supp} \, \mu_k^{(\text{pod})}$ and support boxes B of $\nu_k^{(\text{pod})}$ using (4.7). Accordingly compute $d(W_2 b, W_1(A))$ for each test point $b \in \text{supp} \, \nu_k^{(\text{pod})}$ and support boxes A of $\mu_k^{(\text{pod})}$.

3. For given $\varepsilon > 0$, compute $\mathbf{v}_j(\varepsilon)$ as an approximation of $\nu_k^{(\text{pod})}(W_1(A_j)^\varepsilon)$ by counting all test points for which the corresponding distance is smaller than ε as in (4.6). Analogously, compute $\mathbf{u}_i(\varepsilon)$ as an approximation of $\mu_k^{(\text{pod})}(W_2(B_i)^\varepsilon)$.

4. Step 3 allows the evaluation of the antitone scalar functions $f_i : \mathbb{R}_+ \to \mathbb{R}$ given by (4.8). Use a root finder (e.g. quasi-Newton, bisection) to find the root ε^* of $g(\varepsilon) = \max(f_1(\varepsilon), f_2(\varepsilon))$. Choose this value as an approximation of the Prohorov distance:

$$\varepsilon^* \approx p(\mu_k^{(\text{pod})}, \nu_k^{(\text{pod})}).$$

We recall that distances are computed only in the low-dimensional POD spaces during the algorithm. Hence, the algorithm is quite efficient as we will see in Section 4.1.5.

4.1.4 An alternative distance using negative Sobolev norms

Another theoretical concept is based on the distributional notion (3.9). If we pass on to the theory of distributions we can find a Hilbert space, more exactly a Sobolev space with negative parameter (cf. Definition A.52), containing the corresponding distributions to measures $\mu_k^{(\text{pod})}$ with thin support.

The following proposition gives a representation of the Fourier transform of an arbitrary discrete measure in the distributional sense. For details on distributions, see Section A.5 in the appendix. Recall that by Proposition 3.2 a discrete measure $\mu_k^{(\text{pod})}$ given by the POD algorithms 3.1 and 3.3 can be written as

$$\mu_k^{(\text{pod})} = \left(\mu^{(\text{red})} \times \delta_0^{N-\ell}\right) \circ W^T \tag{4.9}$$

with $W \in \mathbb{R}^{N,N}$ orthogonal and $\delta_0^{N-\ell}$ the Dirac measure in $0 \in \mathbb{R}^{N-\ell}$. The low-dimensional discrete measure $\mu_k^{(\text{red})} : \mathcal{B}_k^{(\ell)} \to [0,1]$ is defined by

$$\mu_k^{(\text{red})}(A) = \sum_{i=1}^{K_\nu} u_i \frac{\lambda_\ell(A \cap B_i)}{\lambda_\ell(B_i)} = \int_A h_\ell \, d\lambda_\ell, \quad h_\ell(x) = \sum_{i=1}^{K} \frac{u_i}{\lambda_\ell(B_i)} \mathbb{1}_{B_i}(x),$$

for a given box collection $\mathcal{B}_k^{(\ell)} = \{B_1, \ldots, B_K\}$, $B_i = B(c_i, r_i) \in \mathbb{R}^\ell$, where

$$B(c,r) = \{x \in \mathbb{R}^\ell : |x_i - c_i| \le r_i, i = 1, \ldots, \ell\}. \tag{4.10}$$

Proposition 4.4. *Let* $\mu_k^{(\text{pod})}$ *be a discrete measure given by (4.9). Then the Fourier transform* $\mathcal{F}(\mu_k^{(\text{pod})}) : \mathbb{R}^N \to \mathbb{R}$ *of* $\mu_k^{(\text{pod})} \in \mathcal{S}'$ *can be derived explicitly in terms of the coordinates of the support boxes of* $\mu_k^{(\text{red})}$:

$$\mathcal{F}(\mu_k^{(\text{pod})})(\xi) = \sum_{i=1}^{K} u_i \, h_{c_i, r_i}(W_1^T \xi)$$

where $h_{c,r}(x) = e^{-2\pi i x^T c} \prod_{j=1}^{\ell} \operatorname{sinc}(2r_j x_j)$ *and* $W = (W_1 W_2)$ *with* $W_1 \in \mathbb{R}^{N,\ell}$.

Proof. By (A.22), (A.23), (A.24) and the linearity of $\mathcal{F}$, the Fourier transform of $\mu_k^{(\text{pod})}$ as above is given by

$$\mathcal{F}(\mu_k^{(\text{pod})})(\xi) \overset{(\text{A.22})}{=} \left[|\det W^T|^{-1} \mathcal{F}(\mu_k^{(\text{red})} \times \delta_0^{N-\ell}) \circ (W)^{-1}\right](\xi)$$

$$= \left[\mathcal{F}(\mu_k^{(\text{red})} \times \delta_0^{N-\ell}) \circ W^T\right](\xi) \overset{(\text{A.23})}{=} \mathcal{F}(\mu_k^{(\text{red})})(W_1^T \xi) \widehat{\delta_0^{N-\ell}}(W_2^T \xi)$$

$$\overset{(\text{A.24})}{=} \mathcal{F}(\mu_k^{(\text{red})})(W_1^T \xi) = \sum_{i=1}^{K} \frac{u_i}{\lambda_\ell(B_i)} \widehat{\mathbb{1}_{B_i}}(W_1^T \xi). \tag{4.11}$$

Observe that with $D = \operatorname{diag}(2r) \in GL(\ell, \mathbb{R})$, the characteristic function $\mathbb{1}_B$ of a box $B = B(c,r)$ as in (4.10) can be written as

$$\mathbb{1}_B = \tau_c \left(\prod_{j=1}^{\ell} \mathbb{1}_{[-\frac{1}{2}, \frac{1}{2}]}\right) \circ D^{-1}$$

with the shift operator τ_y defined in Theorem A.45. By (A.21), (A.22), Theorem A.45 g) and (A.24) we have

$$
\widehat{\mathbb{1}_B}(x) \overset{\substack{(A.21),\\(A.22)}}{=} e^{-2\pi i x^T c} |\det D| \mathcal{F}\Big(\prod_{j=1}^{\ell} \mathbb{1}_{[-\frac{1}{2},\frac{1}{2}]} \Big)(D\,x)
$$

$$
\overset{(\text{Thm.}A.45\text{g})}{=} e^{-2\pi i x^T c} |\det D| \prod_{j=1}^{\ell} \mathcal{F}\big(\mathbb{1}_{[-\frac{1}{2},\frac{1}{2}]} \big)(2r_j x_j)
$$

$$
\overset{(A.24)}{=} e^{-2\pi i x^T c} |\det D| \prod_{j=1}^{\ell} \operatorname{sinc}(2r_j x_j)
$$

$$
= |\det D| h_{c,r}(x). \tag{4.12}
$$

Together with $\det D = \prod_{j=1}^{\ell} 2r_j = \lambda_\ell(B)$, equalities (4.11) and (4.12) result in the statement.   □

Observe that the discrete measures $\mu_k^{(\mathrm{pod})}$ and $\nu_k^{(\mathrm{pod})}$ in $\mathbb{R}^N$ with supports in ℓ_1- and ℓ_2-dimensional subspaces, respectively, are elements of the Sobolev space $H_s(\mathbb{R}^N)$ with index $s < 0$ where

$$
-s \geq \frac{N}{2} > \frac{N - \min(\ell_1, \ell_2)}{2}. \tag{4.13}
$$

Hence we can measure the distance of discrete measures in the negative Sobolev norm (see Definition A.52) by

$$
\big\| \mu_k^{(\mathrm{pod})} - \nu_k^{(\mathrm{pod})} \big\|_{(-N/2)} = \Big(\int h_{\mu_k^{(\mathrm{pod})},\nu_k^{(\mathrm{pod})}}^{(-N/2)}(\xi)\,d\xi \Big)^{1/2} \tag{4.14}
$$

where $h_{\mu,\nu}^{(s)}(\xi) = |\hat\mu - \hat\nu|^2 (1 + |\xi|^2)^s$. As an example, we illustrate the case $N = 2$, $\ell_1 = 1$, $\ell_2 = 2$ and—for simplicity—$W = I_2$ with only one support box for each measure. In detail, we study the measures given in the notation of the PODAIM algorithm 3.1 by

$$
\mu := \mu_k^{(\mathrm{pod})} = \mu_k^{(\mathrm{red})} \times \delta_0^1,
$$

$$
\mu_k^{(\mathrm{red})}(A) = \frac{\lambda_1(B(0,2) \cap A)}{\lambda_1(B(0,2))} = \frac{1}{4}\lambda_1(B(0,2) \cap A),
$$

$$
\nu_a := \nu_k^{(\mathrm{pod})} = \nu_k^{(\mathrm{red})},
$$

$$
\nu_k^{(\mathrm{red})}(A) = \frac{\lambda_2(B\left(0,(2,a)^T\right) \cap A)}{\lambda_2(B\left(0,(2,a)^T\right))} = \frac{1}{8a}\lambda_2(B\left(0,(2,a)^T\right) \cap A)
$$

with the box notation $B(c, r)$ as in (4.10). From the theory and the geometry of the measures, we expect that $\|\mu_2 - \nu_{2,a}\|_{(-1)}$ converges to zero for $a \to 0$. In Figure 4.2 we have plotted the integrand $h_{\mu_2,\nu_{2,a}}^{(1)}$ on $[-2, 2]^2$ for different choices of $a > 0$. The plots show how the integrand converges to zero as the parameter a decreases. In this simple two-dimensional example we can approximate the integral with high accuracy at least on a suitable finite rectangle by a quadrature formula. In Table 4.1 we see indeed a nice convergence behavior of the approximated negative Sobolev norm.

a	2^0	2^{-1}	2^{-2}	2^{-3}	2^{-4}	2^{-5}	2^{-6}	2^{-7}	2^{-8}
sob(a)	0.6220	0.4839	0.3204	0.1827	0.0953	0.0481	0.0239	0.0117	0.0056

Table 4.1: Approximation sob(a) of the Sobolev Norm $\|\mu - \nu_a\|_{(-1)}$ for different choices of parameter by integration with MATLAB's Quadrature Algorithm `quad2d` on $[-\frac{1}{2}, \frac{1}{2}] \times [-1000, 1000]$.

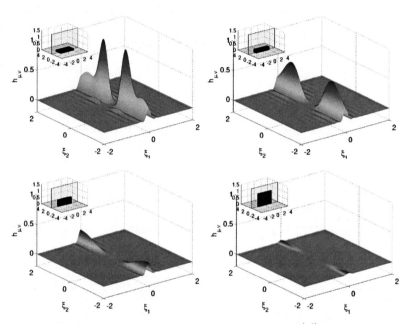

Figure 4.2: From upper left to lower right the Sobolev integrand $h_{\mu,\nu_a}^{(-1)}$ is plotted for the choices $a = 2^0, 2^{-1}, 2^{-2}, 2^{-3}$. The corresponding discrete measure ν_a is displayed in blue in the small figure. The hyperplane containing supp μ is displayed in ambient red.

4.1.5   Prohorov metric vs. Sobolev norm: Numerical realization

In more realistic situations which we will analyze in Chapter 6, the state space dimension N is large. Then the approximation of the N-dimensional integral (4.14) causes difficulties. We show the advantages of the Prohorov approach compared to the approximation of a Sobolev norm in a high-dimensional system in the following.

We consider a discrete dynamical system given by a spatial discretization of the Chafee-Infante problem defined in Chapter 6, using $N = 20$ finite elements (see Chapter 6 for details of the discretization). Using the PODAIM algorithm 3.1, we compute two sequences of discrete measures $\mu_k^{(\text{pod})}$ and $\nu_k^{(\text{pod})}$ given by (3.6). For $\mu_k^{(\text{pod})}$ the number of POD modes was given by $\ell_1 = 2$ while $\ell_2 = 3$ POD modes were used for the generation of $\nu_k^{(\text{pod})}$. We

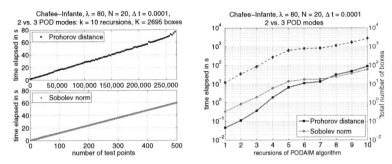

Figure 4.3: Dependencies of the computational times from the number of test points and the number of support boxes K_k of the compared measures. In the left plot, the number of test points increases for the same distance computation $d\big(\mu_{10}^{(\text{pod})}, \nu_{10}^{(\text{pod})}\big)$. In the right plot, for different box collections, the Prohorov distance is computed with K_k times $P_1 = 100$ test points and the Sobolev norm with $P_2 = 500$ test points.

focus our tests on two pairs of measures: the coarse measures $\mu_2^{(\text{pod})}$ and $\nu_2^{(\text{pod})}$ with 16 and 18 support boxes of large scale, i.e. the radius is greater or equal to 1, and two finer measures, $\mu_{10}^{(\text{pod})}$ with $K = 239$ and $\nu_{10}^{(\text{pod})}$ with $K = 2456$ support boxes of smaller scale, i.e. the smallest support boxes are of radius 2^{-8}.

We test our algorithms by computing the distance $d\big(\mu_k^{(\text{pod})}, \nu_k^{(\text{pod})}\big)$. For the Prohorov metric, we use Algorithm 4.3 with a fixed number of P_1 test points per support box, i.e. a total number of $K \cdot P_1$ test points. For the Sobolev norm we have to approximate the 20-dimensional integral given by (4.14). Since the index $s = 10$ given by (4.13) is large we expect no relevant support of the integrand outside the unit cube. Therefore we use a Monte-Carlo approach with P_2 test points in $[-1, 1]^{20}$ and approximate the integral over $[-1, 1]^{20}$ in (4.14) by the following sum

$$\big\| \mu_k^{(\text{pod})} - \nu_k^{(\text{pod})} \big\|_{(-10)} = \int h^{(10)}_{\mu_k^{(\text{pod})}, \nu_k^{(\text{pod})}} \, d\lambda_{20}$$

$$\approx \frac{\lambda_{20}([-1,1]^{20})}{P_2} \sum_{p=1}^{P_2} h^{(10)}_{\mu_k^{(\text{pod})}, \nu_k^{(\text{pod})}}(\xi_p)$$

$$= \frac{2^{20}}{P_2} \sum_{p=1}^{P_2} h^{(10)}_{\mu_k^{(\text{pod})}, \nu_k^{(\text{pod})}}(\xi_p). \tag{4.15}$$

In Figure 4.3 we see how the costs for the computations behave when parameters change. For this, we have measured the time elapsed for the computation of the distances on a standard PC using MATLAB code. In both approximation algorithms the costs grow almost linear with the number of test points. For a fixed number of test points, also the costs grow almost linear with the number of support boxes. It is obvious that this cardinality reflects the complexity of the discrete measures.

Note that for the Prohorov metric, the total number of test points in Figure 4.3 increases up to $2\,695 \cdot 100 = 269\,500$ whereas for the Sobolev norm, we only consider $P_2 \leq 500$ test

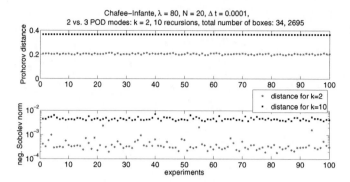

Figure 4.4: Computation of the distance notions for 100 different choices of test points. For the Sobolev norm, the Monte-Carlo approach is based on $P_2 = 10,000$ test points, for the Prohorov metric $P_1 = 100$ test points per box were chosen. Results for $d\big(\mu_2^{(\text{pod})}, \nu_2^{(\text{pod})}\big)$ in red and $d\big(\mu_{10}^{(\text{pod})}, \nu_{10}^{(\text{pod})}\big)$ in blue.

points. Nevertheless the Prohorov norm is computed very efficiently due to the fact that the distance computations are made in low-dimensional spaces (here $\ell_1 = 2$ and $\ell_2 = 3$), cf. (4.7).

Although the algorithm for the computation of the Sobolev norm seems to be quite efficient at first sight, in Figure 4.4 we show that the discretization (4.15) does not yield stable results in our example. The reason is that the number of test points for the Monte-Carlo approach is much too small to provide good approximation results for the 20-dimensional integral. This is illustrated by the comparison of results for the computation of $d\big(\mu_2^{(\text{pod})}, \nu_2^{(\text{pod})}\big)$ and $d\big(\mu_{10}^{(\text{pod})}, \nu_{10}^{(\text{pod})}\big)$ as explained above. We repeat 100 random experiments and show the dependency of the results from the choice of test points in Figure 4.4. We see that the Prohorov metric is quite stable if we use $P_1 = 100$ test points. When comparing the finer measures $\mu_{10}^{(\text{pod})}$, $\nu_{10}^{(\text{pod})}$, the influence of the distribution of test points vanishes almost completely. On the other hand, even though we use a quite large number of test points $P_2 = 10\,000$, the approximation experiments for the Sobolev norm present a large variance. This indicates that the true value is approximated very badly.

A closer look on Figure 4.4 shows that the dependency from the choice of test points obviously decreases with growing complexity of measures. Numerical experiments indeed show that for $k = 2$, the number of test points per box has a big influence on the stability of the approximation result while for $k = 10$, even one test point per box suffices for a stable result as in Figure 4.4.

Remark. *The Monte-Carlo approach for the computation of the negative Sobolev norm can certainly be improved a lot by passing over to a smart choice of a grid. We refer to the theory of sparse grid approximations (see [GK09]) for an approach coming from the theory of finite elements. Nevertheless we doubt to get as stable results in even higher dimensions as with the Prohorov metric.*

4.2 Contraction results for the Frobenius-Perron operator

4.2.1 Contraction arguments for discrete measures

In Chapter 1 we presented the convergence result by Dellnitz, Junge and co-workers concerning the measures μ_k and μ_{SRB}. There, μ_k were discrete measures defined as fixed points of the finite Perron-Frobenius matrix $P_k \in \mathbb{R}^{K,K}$, $K = \#\mathcal{B}$ and μ_{SRB} was an SRB measure of the underlying discrete dynamical system. Roughly speaking, the result was of the form

$$\lim_{\varepsilon \to 0} \lim_{k \to \infty} \mu_k(\varepsilon) = \mu_{SRB} \tag{4.16}$$

where the original dynamical system was replaced by stochastically perturbed systems. In particular, no direct proof of the statement $\lim_{k \to \infty} \mu_k = \mu_{SRB}$ was given. Such a result is not known from the literature to our knowledge. The convergence (4.16) is based on compactness properties of the corresponding Frobenius-Perron operator P_ε in the stochastic setting. This implies that there is no positive answer to the following question: Given a tolerance tol > 0 and a perturbation parameter $\varepsilon > 0$, how fine should we choose the box collection such that the discrete measure is closer to μ_{SRB} as the given tolerance? In other words, does there exist a $K = K(\varepsilon, \text{tol})$ with

$$d(\mu_k(\varepsilon), \mu_{SRB}) < \text{tol} \quad \text{for all } k \geq K$$

where the distance is measured in a suitable metric?

It would be desirable to have such convergence rates or, even better, a direct convergence result

$$d(\mu_k, \mu_{SRB}) \xrightarrow{k \to \infty} 0 \tag{4.17}$$

in a suitable metric. Such rates could result from a contraction property of the Frobenius-Perron operator $P : \mathcal{M}^1(\mathbb{R}^N) \to \mathcal{M}^1(\mathbb{R}^N)$ of the original system that is consistent with a contraction property of the discrete Perron-Frobenius matrix . Though we are far from being capable of showing such a result, and it is doubtful that it holds in all generality, it is worthwhile to study some steps towards such a convergence theory.

Therefore we have a look at the fully discretized system, i.e. we study the Perron-Frobenius matrix acting on discrete measures on a fixed box collection. Asking for contraction in a suitable metric, we present the following results:

- It is well-known that contraction in a proper norm $\| \cdot \|_{P,\varepsilon}$ can be achieved where $\| \cdot \|_{P,\varepsilon}$ strongly depends on the full spectrum of P. Further, contraction in the Hilbert metric is also known to hold for positive Perron-Frobenius matrices.

- Under certain regularity assumptions on P, we will show contraction in a slightly modified discretization of the Prohorov metric. Recall that the Prohorov metric is a suitable distance notion for both probability measures on the original state space and discrete measures. We show in an example that we cannot omit the regularity assumptions in general. A modified Prohorov metric $\hat{p}$ does not exist such that every positive Perron-Frobenius matrix is contracting in $\hat{p}$.

These considerations may give some insight into the chances and limitations of contraction-based convergence results. We will pursue the convergence analysis further in Chapter 5.

Now we collect some results concerning the contraction of the Frobenius-Perron operator in the discrete case. We recall the setting of discrete measures defined in the AIM algorithm 1.30. Let a box collection in a positive invariant set $B_0^{(N)} \subset \mathbb{R}^N$ be given by $\mathcal{B} = \{B_1, \ldots, B_K\}$. We denote by

$$\Delta^{K-1} = \{u \in \mathbb{R}^K : \mathbb{1}^T u = 1, u \geq 0\}$$

the standard simplex in $\mathbb{R}^K$. Every discrete measure μ on $\mathcal{B}$ is represented by a vector $u \in \Delta^{K-1} \subset \mathbb{R}^K$ via

$$\mu(A) = \sum_{i=1}^{K} u_i \frac{\lambda_N(A \cap B_i)}{\lambda_N(B_i)}. \tag{4.18}$$

In the following, we identify the discrete measure μ with its defining vector $u \in \Delta^{K-1}$. The corresponding Frobenius-Perron operator is given by the matrix $P = (p_{ij})_{ij} \in \mathbb{R}^{K,K}$ defined by

$$p_{ij} = \frac{\lambda_N(B_j \cap F^{-1}(B_i))}{\lambda_N(B_j)}, \quad 1 \leq i, j \leq K.$$

Observe that since $B_0^{(N)}$ is positive invariant, the Perron-Frobenius matrix P satisfies

$$\mathbb{1}^T P = \mathbb{1}^T \text{ and } P \geq 0, \tag{4.19}$$

hence P leaves Δ^{K-1} invariant.

Contraction in some induced vector norm

As a first observation we cite the well-known result that, under the regularity assumption that P is primitive, we can always find a norm defined by the spectral properties of P such that P is contracting in this norm. We give the proof as a motivation of what follows.

Proposition 4.5. *Let the Perron-Frobenius matrix $P \in \mathbb{R}^{K,K}$ be primitive, i.e. $P^m > 0$ for an $m \in \mathbb{N}$. Let $\sigma_{red}(P) = \sigma(P) \setminus \{1\}$ denote the reduced spectrum of P with corresponding spectral radius $\rho_{red}(P) < 1$. For any*

$$0 < \varepsilon < 1 - \rho_{red}(P)$$

there exists a norm $\| \cdot \|_{P,\varepsilon}$ on $\mathbb{R}^K$ such that P is contracting on Δ^{K-1} with constant $L_{P,\varepsilon} := \rho_{red}(P) + \varepsilon < 1$,

$$\|P\mu - P\nu\|_{P,\varepsilon} \leq L_{P,\varepsilon} \|\mu - \nu\|_{P,\varepsilon} \quad \text{for all } \mu, \nu \in \Delta^{K-1}. \tag{4.20}$$

Proof. For the stochastic primitive matrix $P \in \mathbb{R}^{K,K}$, the Perron-Frobenius Theorem A.2 implies that P has a simple eigenvalue 1 and no other eigenvalue λ exists with $|\lambda| = 1$. Hence, indeed $\rho_{red}(P) < 1$.

We denote the nonnegative Perron eigenvector by $\bar{\mu} \in \Delta^{K-1}$. Consider the Jordan decomposition of P with eigenvalues $1 > |\lambda_1| \geq \ldots \geq |\lambda_k|$, $\lambda_i \in \mathbb{C}$:

$$U^{-1}PU = J = \operatorname{diag}(1, J_2, \ldots, J_k), \tag{4.21}$$

where $U = \operatorname{col}(\bar{\mu}, u_2, \ldots, u_K)$ is invertible, $J_i = \lambda_i I_{m_i} + E_{m_i}$, $i = 1, \ldots, k$ with the nilpotent matrix $E_{m_i} = (\delta_{m+1,n})_{mn}$ and $K = \sum_{i=1}^{k} m_i$. Note that $\lim_{\ell \to \infty} J_i^\ell = 0$ since $|\lambda_i| < 1$, $i = 2, \ldots, k$. With $\mu \in \Delta^{K-1}$ and $\alpha = U^{-1}\mu$ we get

$$P^\ell \mu = P^\ell U \alpha = U J^\ell \alpha = U \operatorname{diag}(1, J_2^\ell, \ldots, J_k^\ell) \alpha \xrightarrow{\ell \to \infty} U \operatorname{diag}(1, 0, \ldots, 0) \alpha = \alpha_1 \bar{\mu}$$

and it follows
$$\alpha_1 = \mathbb{1}^T \alpha_1 \bar{\mu} = \lim_{\ell \to \infty} \mathbb{1}^T P^\ell \mu \overset{(4.19)}{=} 1.$$

Hence, every $\mu \in \Delta^{K-1}$ is represented by $\mu = U(1, \alpha_2, \ldots, \alpha_K)^T$.

Now let $0 < \varepsilon < 1 - \rho_{\text{red}}(P)$ be given. With $D = \text{diag}(1, D_2, \ldots, D_k)$, where $D_i = \text{diag}(\varepsilon, \ldots, \varepsilon^{m_i})$, $i = 2, \ldots, k$ and $S := UD$ we define

$$\tilde{J} := S^{-1} P S = \text{diag}(1, \tilde{J}_2, \ldots, \tilde{J}_k), \text{ where } \tilde{J}_i = \lambda_i I_{m_i} + \varepsilon E_{m_i}. \tag{4.22}$$

We define the norm $\| \cdot \|_{P,\varepsilon}$ by
$$\|v\|_{P,\varepsilon} = \|S^{-1} v\|_\infty.$$

In this norm, for every $\mu = U\alpha, \nu = U\beta \in \Delta^{K-1}$ we get the representations:

$$\|\mu - \nu\|_{P,\varepsilon} = \|S^{-1} U(\alpha - \beta)\|_\infty = \|w\|_\infty,$$

where $w = D^{-1}(\alpha - \beta) = (0, w_2, \ldots, w_K)^T$ and

$$\|P\mu - P\nu\|_{P,\varepsilon} = \|S^{-1} P U(\alpha - \beta)\|_\infty = \|D^{-1} J(\alpha - \beta)\|_\infty = \|D^{-1} J D w\|_\infty = \|\tilde{J} w\|_\infty.$$

Since the first entry of w vanishes, the contraction property follows:

$$\|P\mu - P\nu\|_{P,\varepsilon} = \|\text{diag}(\tilde{J}_2, \ldots, \tilde{J}_k)(w_2, \ldots, w_K)^T\|_\infty$$
$$\leq \|\text{diag}(\tilde{J}_2, \ldots, \tilde{J}_k)\|_\infty \|w\|_\infty \leq (|\lambda_2| + \varepsilon)\|\mu - \nu\|_{P,\varepsilon} = L_{P,\varepsilon} \|\mu - \nu\|_{P,\varepsilon}.$$

$\square$

The standard result of contraction in the P-induced norm as defined above is not satisfying in our context since the norm itself depends on the whole spectrum of P. Hence, it depends strongly on the box collection defining the discretization of the continuous Frobenius-Perron operator.

Contraction in the Hilbert metric

It is well-known that a suitable distance notion on the standard simplex is given by the Hilbert geometry. For details of the construction see [Pap05], Section 5.6. For a short introduction we denote the *cross ratio* of points $a_1, a_2, b_1, b_2 \in \mathbb{R}^K$ with $a_1 \neq b_1$, $a_2 \neq b_2$ by

$$[a_1, a_2, b_1, b_2] := \frac{\|a_2 - b_1\|_2 \|a_1 - b_2\|_2}{\|a_1 - b_1\|_2 \|a_2 - b_2\|_2}.$$

On a bounded and open convex set $C \subset \mathbb{R}^K$ it is easy to show that for $a_1, a_2 \in C$ the line $l(a_1, a_2)$ containing them has exactly two intersections with ∂C in $b_1, b_2 \in \partial C$. Assume b_1, a_1, a_2, b_2 are aligned in that order on $l(a_1, a_2)$. Then the operator $h_C : C \times C \to \mathbb{R}_+$ is defined by

$$h_C(a_1, a_2) = \begin{cases} \ln[a_1, a_2, b_1, b_2] & \text{if } a_1 \neq a_2, \\ 0 & \text{if } a_1 = a_2 \end{cases}$$

By the properties of the cross ratios it follows that h_C defines a metric, called the *Hilbert metric* or the *projective metric* on C. Hence we can define the Hilbert metric $h = h_{\text{int}(\Delta^{K-1})}$

on the interior $\text{int}(\Delta^{K-1})$ of the standard simplex. Observe that for a positive Perron-Frobenius matrix P

$$P(\text{int}(\Delta^{K-1})) \subset \text{int}(\Delta^{K-1}).$$

In [Bir57] Birkhoff already proved that every positive matrix is contracting in the Hilbert metric. In this way one obtains an alternative proof of the Perron-Frobenius Theorem A.2 for positive matrices. There is a whole theory concerning implications and generalizations of this early result, see [Kra86], [Kra01], [NVL99]. We mention the result by Kohlberg and Pratt [KP82] stating that the Hilbert metric is—up to scaling—the only metric that guarantees contraction of every linear transformation of the positive cone (i.e. every positive matrix in the finite-dimensional case).

Despite these nice contraction results in the Hilbert metric, this approach seems not to be suitable for convergence of discrete measures due to the lack of a natural generalization to a metric on $\mathcal{M}^1(\mathbb{R}^N)$. Therefore, we continue our analysis with the case of the Prohorov metric p. This metric is suitable for both the space $\mathcal{M}^1(\mathbb{R}^N)$ and the space of discrete measures where p is discretized along support box collections.

4.2.2 Contraction in metrics of Prohorov type

As discussed before, the Prohorov metric has a suitable numerical realization for discrete measures. We analyze this discretization in the following. For simplicity, we restrict ourselves to the one-dimensional case, i.e. $B_0^{(N)} = [-1, 1]$ with an equidistant box collection

$$\mathcal{B} = \{B_1, \ldots, B_K\}, \ B_i = B\left(-1 + \frac{2i-1}{K}, \frac{1}{K} \right), \quad i = 1, \ldots, K.$$

We define an abstract distance notion for measures $\mu, \nu \in \Delta^{K-1}$ on $\mathcal{B}$.

Definition 4.6. For given $w \in \mathbb{R}^K$, $A \in \mathbb{R}^{K,K}$ with $A, w \geq 0$ we define a *distance* $p : \Delta^{K-1} \times \Delta^{K-1} \to \mathbb{R}_+$ of *Prohorov type* by

$$p(\mu, \nu) = \min \left\{ \varepsilon > 0 : \begin{array}{l} \mu - \nu \leq \varepsilon(w + A\nu), \\ \nu - \mu \leq \varepsilon(w + A\mu) \end{array} \right\}. \tag{4.23}$$

We show that the discretized Prohorov metric $\tilde{p}$ as defined in (4.5) is a distance of Prohorov type in the sense of Definition 4.6. Recall that $\tilde{p}(\mu, \nu)$ is given by

$$\tilde{p}(\mu, \nu) = \min\{\varepsilon > 0 : u_i \leq \nu(B_i^\varepsilon) + \varepsilon, v_i \leq \mu(B_i^\varepsilon) + \varepsilon \text{ for all } i = 1, \ldots, K\}. \tag{4.24}$$

For $\varepsilon \leq \frac{2}{K}$ we get $B_i^\varepsilon \subset B_{i-1} \cup B_i \cup B_{i+1}$ where $B_0 = B_{K+1} = \emptyset$ and accordingly

$$\mu(B_i^\varepsilon) = u_i + \varepsilon \frac{K}{2}(u_{i-1} + u_{i+1}), \text{ where } u_0 = u_{K+1} = 0. \tag{4.25}$$

For $\tilde{p}(\mu, \nu) \leq \frac{2}{K}$, $\tilde{p}$ can be written as

$$\tilde{p}(\mu, \nu) = \min\{\varepsilon > 0 : \mu - \nu \leq \varepsilon\left(\mathbb{1} + \frac{K}{2}R\nu\right), \ \nu - \mu \leq \varepsilon\left(\mathbb{1} + \frac{K}{2}R\mu\right)\} \tag{4.26}$$

where the neighborhood matrix $R \in \mathbb{R}^{K,K}$ is given by

$$R = \begin{pmatrix} 0 & 1 & & & 0 \\ 1 & \ddots & \ddots & & \\ & \ddots & \ddots & 1 \\ 0 & & 1 & 0 \end{pmatrix}. \tag{4.27}$$

Hence $\tilde{p}$ is a distance of Prohorov type with $w = \mathbb{1}$ and $A = \frac{K}{2}R$. However, $\tilde{p}$ defined by (4.26) with R given by (4.27) is not a metric on a subset of Δ^{K-1} since the triangle inequality does not hold. As an example, choose Δ^3 and for $\delta < \frac{1}{4}$ consider the discrete measures

$$\omega = \frac{1}{4}\mathbb{1}, \quad \mu_\delta = \omega + \delta(1, -1, -1, 1)^T,$$

$$\nu_\delta = \omega + \delta(1, 1, -1, -1)^T \in \Delta^3.$$

Note that the distance in the maximum norm between each of these measures is arbitrarily small for small $\delta > 0$. Nevertheless, an elementary calculation shows

$$\tilde{p}(\mu_\delta, \omega) + \tilde{p}(\omega, \nu_\delta) = \frac{2}{3}\delta + \frac{2}{3}\delta = \frac{4}{3}\delta < \frac{4}{3 - 4\delta}\delta = \tilde{p}(\mu_\delta, \nu_\delta).$$

Hence, the triangle inequality is violated.

Therefore we consider other neighborhood matrices $A \in \{\frac{K}{2}\tilde{R}, \frac{K}{2}S\}$, where $\tilde{R}, S \in \mathbb{R}^{K,K}$ are defined by

$$\tilde{R} = R + \begin{pmatrix} 0 & \cdots & 0 & 1 \\ \vdots & \ddots & & 0 \\ 0 & & \ddots & \vdots \\ 1 & 0 & \cdots & 0 \end{pmatrix}, \quad S = (1 - \delta_{ij})_{ij} = \begin{pmatrix} 0 & 1 & \cdots & 1 \\ 1 & \ddots & \ddots & \vdots \\ \vdots & \ddots & \ddots & 1 \\ 1 & \cdots & 1 & 0 \end{pmatrix}.$$

The slight modification from R to the stochastic matrix $\tilde{R}$ means that the boxes B_1 and B_K are viewed as neighbors in our setting. However this modification does not suffice. The distance of Prohorov type with $w = \mathbb{1}$ and $A = \frac{K}{2}\tilde{R}$ does not satisfy the triangle inequality as a $K = 5$-dimensional example shows.

The definition of $\bar{p} : \Delta^{K-1} \times \Delta^{K-1} \to \mathbb{R}_+$ as a distance of Prohorov type with $w = \mathbb{1}$ and $A = \frac{K}{2}S$ leads to a metric on a proper subset of Δ^{K-1}:

Proposition 4.7. *The operator $\bar{p}$ defines a metric on every subset U of Δ^{K-1} with*

$$\operatorname{diam}_{\bar{p}}(U) := \sup_{\mu,\nu \in U} \bar{p}(\mu, \nu) \leq \frac{2}{K}.$$

The discrete Prohorov metric defined by $\bar{p}$ is equivalent to the metric induced by the maximum norm on $\mathbb{1}^\perp := \{v \in \mathbb{R}^K : \mathbb{1}^T v = 0\}$:

$$\frac{2}{2 + K}\|\mu - \nu\|_\infty \leq \bar{p}(\mu, \nu) \leq \|\mu - \nu\|_\infty \quad \text{for all } \mu, \nu \in \Delta^{K-1}, p(\mu, \nu) \leq \frac{2}{K}.$$

However, the discrete Prohorov metric is not induced by a norm on $\mathbb{1}^\perp$.

Proof. Recall the definition of $\bar{p} : \Delta^{K-1} \times \Delta^{K-1} \to \mathbb{R}_+$,

$$\bar{p}(\mu, \nu) = \min\{\varepsilon > 0 : \mu - \nu \leq \varepsilon\left(\mathbb{1} + \frac{K}{2}S\nu\right), \nu - \mu \leq \varepsilon\left(\mathbb{1} + \frac{K}{2}S\mu\right)\} \quad (4.28)$$

with $s_{ij} = 1 - \delta_{ij}$. By definition, $\bar{p}$ is symmetric and maps into $[0, 1] \subset \mathbb{R}_+$. Furthermore, $\bar{p}(\mu, \nu) = 0$ implies $\mu - \nu, \nu - \mu \leq 0$, hence $\mu = \nu$. Thus $\bar{p}$ is positive definite.

The crucial point is the triangle inequality as we have seen in the preceding remark. Therefore observe that by definition of S we can reformulate the inequalities in (4.28) to

$$\mu - \nu \le \varepsilon\big(1 + \frac{K}{2}S\nu\big) \iff u_i - v_i \le \varepsilon\big(1 + \frac{K}{2}(1 - v_i)\big), \qquad i = 1, \dots, K \qquad (4.29)$$

$$\iff u_i \le \big(1 + \frac{K}{2}\big)\varepsilon + \big(1 - \frac{K}{2}\varepsilon\big)v_i, \quad i = 1, \dots, K$$

Hence we can write $\bar{p}$ as

$$\bar{p}(\mu, \nu) = \min\{\varepsilon > 0 : \begin{array}{l} u_i \le (1 + \frac{K}{2})\varepsilon + (1 - \frac{K}{2}\varepsilon)v_i, \\ v_i \le (1 + \frac{K}{2})\varepsilon + (1 - \frac{K}{2}\varepsilon)u_i, \end{array} i = 1, \dots, K\}. \qquad (4.30)$$

Now let $\mu, \nu, \omega \in \Delta^{K-1}$ be given with $p := \bar{p}(\mu, \nu)$, $q := \bar{p}(\nu, \omega) \le \frac{2}{K}$. It follows

$$w_i \le \big(1 + \frac{K}{2}\big)q + \big(1 - \frac{K}{2}q\big)v_i \le \big(1 + \frac{K}{2}\big)q + \big(1 - \frac{K}{2}q\big)\big[\big(1 + \frac{K}{2}\big)p + \big(1 - \frac{K}{2}p\big)u_i\big]$$

$$= \big(1 + \frac{K}{2}\big)(p + q) - \frac{K}{2}\big(1 + \frac{K}{2}\big)pq + \big(1 - \frac{K}{2}q\big)\big(1 - \frac{K}{2}p\big)u_i$$

$$= \big(1 + \frac{K}{2}\big)(p + q) - \frac{K}{2}\big(1 + \frac{K}{2}\big)pq + \big(1 - \frac{K}{2}(p + q)\big)u_i + \frac{K^2}{4}pqu_i$$

$$= \big(1 + \frac{K}{2}\big)(p + q) + \big(1 - \frac{K}{2}(p + q)\big)u_i + \frac{K}{2}pq\underbrace{\big(\frac{K}{2}u_i - 1 - \frac{K}{2}\big)}_{=\frac{K}{2}(u_i - 1) - 1 < 0}.$$

Hence the first K equations of (4.30) hold with $\varepsilon = p + q$. Analogously we get

$$u_i \le \big(1 + \frac{K}{2}\big)(p + q) + \big(1 - \frac{K}{2}(p + q)\big)w_i$$

and hence the desired result $\bar{p}(\mu, \omega) \le p + q$.

For the equivalence observe that

$$\|\mu - \nu\|_\infty = \max\big(\max_{1 \le i \le K}(u_i - v_i), \max_{1 \le i \le K}(v_i - u_i)\big).$$

So on the one hand, we get

$$u_i - v_i \le \|\mu - \nu\|_\infty \le \|\mu - \nu\|_\infty\big(1 + \frac{K}{2}(1 - v_i)\big)$$

and on the other hand

$$u_i - v_i \le \bar{p}(\mu, \nu)\big(1 + \frac{K}{2}\underbrace{(1 - v_i)}_{\le 1}\big) \le \bar{p}(\mu, \nu)\frac{2 + K}{2}.$$

The corresponding estimates for $v_i - u_i$ follow analogously.

Assume $\bar{p}$ is induced by a vector norm $\|\cdot\|$ on $\mathbb{1}^\perp$, i.e.

$$\bar{p}(\mu, \nu) = \|\mu - \nu\| \quad \text{for all } \mu, \nu \in \Delta^{K-1}, \bar{p}(\mu, \nu) \le \frac{2}{K}.$$

For any $\mu \in \Delta^{K-1}$ and $w \in \mathbb{1}^\perp$ with $\bar{p}(\mu, \mu + w) \le \frac{2}{K}$ the norm of w can be expressed in terms of the metric

$$\|w\| = \bar{p}(\mu, \mu + w).$$

In particular for given $\lambda > 0$ with

$$\bar{p}(\mu, \mu + w), \bar{p}(\mu, \mu + \lambda w) \leq \frac{2}{K},$$

the homogeneity of $\|\cdot\|$ implies

$$\bar{p}(\mu, \mu + \lambda w) = \|\lambda w\| \overset{!}{=} \lambda \|w\| = \lambda \bar{p}(\mu, \mu + w). \tag{4.31}$$

We give a counterexample for (4.31) by $\mu = \frac{1}{6}(2, 1, 3)^T$ and $w_\varepsilon = \frac{\varepsilon}{6}(-1, 1, 0)^T$, $\varepsilon > 0$ small. By an easy calculation, we can derive an explicit formula for the Prohorov distance in the case $0 \leq \lambda \varepsilon \leq 2$:

$$\bar{p}(\mu, \mu + \lambda w_\varepsilon) = \frac{2\lambda\varepsilon}{3(8 + \lambda\varepsilon)}$$

Clearly, this expression is not homogeneous in λ.   □

Using the distance $\bar{p}$ of Prohorov type that defines a metric on a subset of the standard simplex, we indeed get contraction on such a subset if we impose further assumptions on the Perron-Frobenius matrix.

Proposition 4.8. *Let the Perron-Frobenius matrix $P \in \mathbb{R}^{K,K}$ be primitive. Let the minimum of row $i \in \{1, \ldots, K\}$ be reached at $1 \leq k_i \leq K$, i.e.*

$$k_i := \arg\min_{1 \leq k \leq K} p_{ik}. \tag{4.32}$$

If P satisfies

$$L_i := p_{i,k_i} + \sum_{\substack{k_i \neq j = 1}}^{K} (p_{ij} - p_{i,k_i}) < 1 \quad \text{for all } 1 \leq i \leq K \tag{4.33}$$

then P is contracting on $K_{\bar{p}}(\bar{\mu}, \frac{1}{K})$ with respect to the discrete Prohorov distance $\bar{p}$:

$$\bar{p}(P\mu, P\nu) \leq L_{P,\bar{p}}\,\bar{p}(\mu, \nu) \quad \text{for all } \mu, \nu \in K_{\bar{p}}(\bar{\mu}, \frac{1}{K}).$$

The contraction constant is given by $L_{P,\bar{p}} := \max_{1 \leq i \leq K} L_i \in (0, 1)$.

Proof. Assume a primitive Perron-Frobenius matrix $P \in \mathbb{R}^{K,K}$ satisfies (4.33) and let k_i be defined by (4.32). Observe that for arbitrary $\mu \in \Delta^{K-1}$

$$(P\mu)_i = \sum_{j=1}^{K} p_{ij} u_j = \sum_{k_i \neq j = 1}^{K} p_{ij} u_j + p_{i,k_i}\Big(1 - \sum_{k_i \neq j = 1}^{K} u_j\Big) = p_{i,k_i} + \sum_{k_i \neq j = 1}^{K} (p_{ij} - p_{i,k_i}) u_j. \tag{4.34}$$

Now let $\mu = (u_1, \ldots, u_K)^T$ and $\nu = (v_1, \ldots, v_K)^T \in \Delta^{K-1}$ be given with $q := \bar{p}(\mu, \nu) \leq \frac{K}{2}$.

We estimate $\bar{p}(P\mu, P\nu)$ as follows

$$(P\mu - P\nu)_i \overset{(4.34)}{=} \sum_{k_i \neq j=1}^{K} (p_{ij} - p_{i,k_i})(u_j - v_j) \overset{(4.29)}{\leq} \sum_{k_i \neq j=1}^{K} (p_{ij} - p_{i,k_i}) q \left(1 + \frac{K}{2} - \frac{K}{2} v_j\right)$$

$$= -q\frac{K}{2} \sum_{k_i \neq j=1}^{K} (p_{ij} - p_{i,k_i}) v_j + q \sum_{k_i \neq j=1}^{K} (p_{ij} - p_{i,k_i}) \left(1 + \frac{K}{2}\right)$$

$$\overset{(4.34)}{=} -q\frac{K}{2}(P\nu)_i + q\frac{K}{2} p_{i,k_i} + q\left(1 + \frac{K}{2}\right) \sum_{k_i \neq j=1}^{K} (p_{ij} - p_{i,k_i})$$

$$\leq -q\frac{K}{2}(P\nu)_i + q\left(1 + \frac{K}{2}\right)\left[p_{i,k_i} + \sum_{k_i \neq j=1}^{K} (p_{ij} - p_{i,k_i})\right]$$

$$\leq L_{P,\bar{p}} q \left(\left(1 + \frac{K}{2}\right) - \frac{K}{2}(P\nu)_i\right)$$

where the last estimate follows from $L_{P,\bar{p}} \in (0,1)$. By interchanging μ and ν in the above expression we get the analogous result for $(P\nu - P\mu)_i$. By definition of the metric $\bar{p}$, see (4.29), we get

$$\bar{p}(P\mu, P\nu) \leq L_{P,\bar{p}} \bar{p}(\mu, \nu).$$

$\square$

Remark. *We note some implications of the proposition, in particular of the contraction condition (4.33) and further considerations arising from it.*

1. The two-dimensional case: *Observe that for $K = 2$ Proposition 4.8 implies the contraction property in $\bar{p}$ for every positive Perron-Frobenius matrix, because we get*

$$(4.33) \iff \min(p_{i1}, p_{i2}) + \max(p_{i1}, p_{i2}) - \min(p_{i1}, p_{i2}) < 1 \quad \text{for } i \in \{1,2\}$$
$$\iff \max(p_{i1}, p_{i2}) < 1 \iff p_{ij} < 1 \quad \text{for } i, j \in \{1,2\}$$
$$\iff p_{ij} > 0 \quad \text{for } i, j \in \{1,2\}$$

since P is stochastic.

2. Examples for (4.33): *We write the condition (4.33) in an equivalent form*

$$\sum_{j=1}^{K} p_{ij} < 1 + (K-1) p_{i,k_i}. \tag{4.35}$$

To get a feeling for condition (4.35), consider two cases for $K = 3$ given by the positive Perron-Frobenius matrices

$$P_1 = \frac{1}{12} \begin{pmatrix} 9 & 10 & 10 \\ 1 & 1 & 1 \\ 2 & 1 & 1 \end{pmatrix}, \quad P_1 = \frac{1}{12} \begin{pmatrix} 1 & 10 & 10 \\ 9 & 1 & 1 \\ 2 & 1 & 1 \end{pmatrix}.$$

P_1 and P_2 differ only by a permutation of the entries p_{11} and p_{21}. Observe that in both cases the first row is the only critical row, since the other row sums are smaller than one. In the first row, condition (4.35) holds for P_1 but not for P_2.

Indeed, P_2 is not contracting as the following example shows. For

$$\mu = \frac{1}{3}\mathbb{1}, \quad \nu = \frac{1}{2}(0,1,1)^T$$

we get for the Prohorov distances

$$\bar{p}(P\mu, P\nu) = \frac{2}{13} > \frac{2}{15} = \bar{p}(\mu, \nu).$$

3. Modified Prohorov distance: *Since condition (4.33) is quite restrictive, one may look for modifications such that we get contraction for an arbitrary Perron-Frobenius matrix P.*

Recall the result of [KP82] stating that a metric that works for every primitive matrix is essentially given by the Hilbert metric. Thus we expect that no metric of Prohorov type exists satisfying contraction for an arbitrary Perron-Frobenius matrix. Indeed, the ansatz of a general distance $\hat{p}$ of Prohorov type (4.23) with $w \in \mathbb{R}^K$ and a matrix $A \in \mathbb{R}^{K,K}$ does not lead to contraction for general Perron-Frobenius matrices:

Assume $\hat{p}$ implies contraction for an arbitrary Perron-Frobenius matrix P, i.e. there exists an $L \in [0,1)$ with

$$\hat{p}(P\mu, P\nu) \leq L\hat{p}(\mu, \nu) \quad \text{for all } \mu, \nu \in \Delta^{K-1}. \tag{4.36}$$

Observe that a sufficient—and more or less substantial—condition for (4.36) is given by the vector inequality

$$P(w + A\mu) \leq L(w + AP\mu) \quad \text{for all } \mu \in \Delta^{K-1}. \tag{4.37}$$

It can easily be shown that even for $L = 1$ the property (4.37) implies for primitive P that A and P are almost commuting, more precisely

$$[A, P] = AP - PA = (Pw - w)\mathbb{1}^T,$$

and additionally that (4.37) holds with equality,

$$P(w + A\mu) = w + AP\mu \quad \text{for all } \mu \in \Delta^{K-1}.$$

In particular, there is no $w \geq 0, A \geq 0$ such that (4.37) holds with $L < 1$ for every Perron-Frobenius matrix P. Hence we cannot expect a Prohorov-like metric $\hat{p}$ such that every primitive Perron-Frobenius matrix is contracting in $\hat{p}$.

Chapter 5

Error estimates for discrete and continuous invariant measures

In this chapter, we analyze in detail the reduced-space algorithms introduced in Chapter 3. We do not set up a complete convergence theory for the discrete measures here since general ergodic propositions about POD systems seem to be out of reach. Instead, we focus on some special instructive cases to develop proper error estimates concerning the behavior of the continuous and discrete measures involved in the theory of the PODAIM algorithm. In Section 5.1, we give an overview of the measures, transfer operators and support sets involved in the algorithm. Then we present known results concerning error estimates and convergence of discrete measures. We give some useful relations between invariant measures in the original state space and in the POD space. Finally, we analyze discrete measures in POD spaces and derive explicit error bounds with respect to the discrete measures computed by the AIM algorithm.

5.1 Overview of convergence results for the PODAIM algorithm

Setting for a convergence theory

As in the previous chapters, we consider a discrete dynamical system on $\mathbb{R}^N$,

$$u_{i+1} = F(u_i), \quad i \in \mathbb{N},$$

where $F : \mathbb{R}^N \to \mathbb{R}^N$ is a diffeomorphism. We aim at convergence results for the PODAIM algorithm 3.1. In particular, we will compare the resulting discrete measures arising in the PODAIM algorithm with the discrete measures of the AIM algorithm 1.30. Therefore we recall the setting of the algorithms. Both algorithms are used to approximate a proper invariant measure, e.g. an SRB measure, $\mu_{\text{inv}} \in \mathcal{M}^1(\mathbb{R}^N)$,

$$\mu_{\text{inv}}(A) = \mu_{\text{inv}}(F^{-1}(A)), \quad A \in \mathcal{B}(\mathbb{R}^N),$$

in the original state space $\mathbb{R}^N$. The details are elaborated in Chapter 1. As needed for algorithm 1.30, we assume the existence of a positive invariant box $B_0^{(N)} \subset \mathbb{R}^N$. Hence, the support of the approximated invariant measure $\operatorname{supp} \mu_{\text{inv}}$ is a subset of $B_0^{(N)}$.

For a POD space given by the orthogonal matrix $W_1 = \mathrm{col}(w_1, \ldots, w_\ell) \in \mathbb{R}^{N,\ell}$ we denote the reduced system function by $F_{\mathrm{red}} : \mathbb{R}^\ell \to \mathbb{R}^\ell$,

$$F_{\mathrm{red}}(\alpha) = W_1^T F(W_1 \alpha), \quad \alpha \in \mathbb{R}^\ell. \tag{5.1}$$

As in algorithm 3.1, we also assume that there exists a positive invariant box $B_0^{(\ell)} \subset \mathbb{R}^\ell$ of the reduced system. Hence there is an F_{red}-invariant measure $\mu^{(\mathrm{red})} \in \mathcal{M}^1(\mathbb{R}^\ell)$,

$$\mu^{(\mathrm{red})}(A) = \mu^{(\mathrm{red})}(F_{\mathrm{red}}^{-1}(A)), \quad A \in \mathcal{B}(\mathbb{R}^\ell),$$

with $\mathrm{supp}\,\mu_{(\mathrm{red})} \subset B_0^{(\ell)}$. Corresponding to $\mu^{(\mathrm{red})}$, we define the extended measure $\mu^{(\mathrm{pod})} \in \mathcal{M}^1(\mathbb{R}^N)$ as we did in the PODAIM algorithm 3.1 for a discrete measure, cf. (3.9), by

$$\mu^{(\mathrm{pod})} = \left(\mu^{(\mathrm{red})} \times \delta_0^{N-\ell}\right) \circ W^T \tag{5.2}$$

where $W = \mathrm{col}(w_1, \ldots, w_N) \in \mathbb{R}^{N,N}$ is the extension of $W_1 = \mathrm{col}(w_1, \ldots, w_\ell) \in \mathbb{R}^{N,\ell}$ to an orthogonal square matrix. In the following we consider fixed W, W_1 and $W_2 = \mathrm{col}(w_{\ell+1}, \ldots, w_N) \in \mathbb{R}^{N-\ell,N}$.

We recall the definition of the Frobenius-Perron operators $P : \mathcal{M}^1(\mathbb{R}^N) \to \mathcal{M}^1(\mathbb{R}^N)$ in the original state space and $P^{(\mathrm{red})} : \mathcal{M}^1(\mathbb{R}^\ell) \to \mathcal{M}^1(\mathbb{R}^\ell)$ in the POD space with

$$P\mu(A) = \mu(F^{-1}(A)), \quad A \in \mathcal{B}(\mathbb{R}^N), \quad P^{(\mathrm{red})}\mu(A) = \mu(F_{\mathrm{red}}^{-1}(A)), \quad A \in \mathcal{B}(\mathbb{R}^\ell).$$

Observe that μ_{inv} and $\mu^{(\mathrm{red})}$ are characterized as fixed points of P and $P^{(\mathrm{red})}$, respectively.

The discrete measures μ_k computed by the AIM algorithm 1.30 and $\mu_k^{(\mathrm{red})}$ computed by the PODAIM algorithm 3.1 are defined on box collections

$$\mathcal{B}_k^{(N)} := \{B_1^{(N)}, \ldots, B_{K_N}^{(N)}\}, \quad B_i^{(N)} = B(c_i, r_i) \subset B_0^{(N)}, \quad 1 \le i \le K_N,$$

$$\mathcal{B}_k^{(\ell)} := \{B_1^{(\ell)}, \ldots, B_{K_\ell}^{(\ell)}\}, \quad B_i^{(\ell)} = B(c_i, r_i) \subset B_0^{(\ell)}, \quad 1 \le i \le K_\ell.$$

The discrete measures $\mu_k \in \mathcal{M}^1(\mathbb{R}^N)$ and $\mu_k^{(\mathrm{red})} \in \mathcal{M}^1(\mathbb{R}^\ell)$ are given by

$$\mu_k(A) = \sum_{i=1}^{K_N} u_i^{(N)} \frac{\lambda_N(B_i^{(N)} \cap A)}{\lambda_N(B_i^{(N)})}, \quad A \in \mathcal{B}(\mathbb{R}^N), \tag{5.3}$$

$$\mu_k^{(\mathrm{red})}(A) = \sum_{i=1}^{K_\ell} u_i^{(\ell)} \frac{\lambda_\ell(B_i^{(\ell)} \cap A)}{\lambda_\ell(B_i^{(\ell)})}, \quad A \in \mathcal{B}(\mathbb{R}^\ell). \tag{5.4}$$

Here, the vectors $u^{(N)} \in \mathbb{R}^{K_N}$ and $u^{(\ell)} \in \mathbb{R}^{K_\ell}$ are fixed points of the transition matrices $P_k = (p_{ij})_{ij}$ and $P_k^{(\mathrm{red})} = (p_{ij}^{(\mathrm{red})})_{ij}$ with entries

$$p_{ij} = \frac{\lambda_N(B_j^{(N)} \cap F^{-1}(B_i^{(N)}))}{\lambda_N(B_j^{(N)})}, \quad 1 \le i,j \le K_N, \tag{5.5}$$

$$p_{ij}^{(\mathrm{red})} = \frac{\lambda_\ell(B_j^{(\ell)} \cap F_{\mathrm{red}}^{-1}(B_i^{(\ell)}))}{\lambda_\ell(B_j^{(\ell)})}, \quad 1 \le i,j \le K_\ell. \tag{5.6}$$

As for the continuous measure, the extension of $\mu_k^{(\mathrm{red})}$ to a measure $\mu_k^{(\mathrm{pod})} \in \mathcal{M}^1(\mathbb{R}^N)$ is given by

$$\mu_k^{(\mathrm{pod})} = \left(\mu_k^{(\mathrm{red})} \times \delta_0^{N-\ell}\right) \circ W^T. \tag{5.7}$$

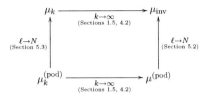

Diagram 5.1: Convergence questions arising in the theory of the PODAIM algorithm 3.1.

Relation to existing results

After having fixed the notation, we categorize the convergence questions arising naturally from the context. See Diagram 5.1 for an overview. We already summarized in Chapter 1 the convergence theory for the AIM algorithm 1.30 developed by Dellnitz and Junge. This refers to the convergence of μ_k to the invariant μ_{inv} in the diagram or, to be more precise, to some SRB measure μ_{SRB}. Note that this approach takes a detour via stochastically perturbed systems.

In Section 4.2 we have shown contraction results of the Frobenius-Perron operator for a fixed box collection. If we assume that (4.33) holds uniformly over all refinement steps and an analogous result holds in an integral form for the continuous transfer operator, this ansatz could also be used for a convergence result. However, such an assumption implies positivity of the transfer operators in general. For positive operators, the convergence theory by Dellnitz and Junge applies. Hence we get no further insight by this approach.

The discrete measures $\mu_k^{(\text{red})}$ are computed by the AIM algorithm in the POD space given by W_1. In principle, the results about convergence of the discrete measures μ_k to an invariant measure μ_{inv} should transfer to results about the convergence of discrete measures $\mu_k^{(\text{red})}$ in the POD space to an F_{red}-invariant measure $\mu^{(\text{red})} \in \mathcal{M}^1(\mathbb{R}^\ell)$. Thereby convergence results for the extended measures $\mu_k^{(\text{pod})}$ and $\mu^{(\text{pod})}$ in $\mathcal{M}^1(\mathbb{R}^N)$ follow.

Another line of thought is as follows. It is reasonable to assume stronger properties for the reduced-order system than for the original system, e.g. F_{red} to be expanding and the invariant measure to be absolutely continuous. For such systems the theory of Liverani, Blank, Keller, Baladi et al. applies, cf. [Liv03], [BKL02], [Bal00]. It states that in this setting the transfer operator satisfies the property of quasi-compactness. Thus some direct convergence results follow for invariant measures in the reduced space.

Our approach

In the following sections we fill in some of the remaining gaps in Diagram 5.1. In Section 5.2, we give positive answers concerning the relation of the continuous measures $\mu^{(\text{pod})}$ with thin support in the POD subspace of $\mathbb{R}^N$ defined by W_1, and the invariant measures μ_{inv} of the original system. To our knowledge the approximation theory for POD models is well-developed only in terms of single trajectories. A general theory of ergodic properties of POD-reduced systems currently seems to be out of scope. In some special cases we work out relations between invariant measures of the original dynamical system and invariant measures in the model-reduced system.

The results for the discrete measures are more comprehensive. Using perturbation theory for Perron-Frobenius matrices, some nice properties of the discrete measures in the original state space and the POD subspace follow. In particular, we give a criterion for when the support boxes of the discrete measures arising in the AIM algorithm each contain a support box of the POD algorithm. Under further assumptions we derive estimates for the distance between the discrete measures in both spaces where the order is given by the thickness of the higher dimensional support boxes. We formulate the main result in Corollary 5.14.

5.2   Relations between the continuous measures

5.2.1   Invariance in original and reduced space

In the following proposition we observe that the model reduction is consistent with respect to ergodic properties in the following sense: Every F_{red}-invariant measure extends to a measure $\mu^{(\text{pod})}$ in the original state space that is invariant under F. Furthermore, all F-invariant measures are extensions of F_{red}-invariant measures, if the POD subspace is uniformly attracting.

Proposition 5.1. *Let a discrete dynamical system be defined by a diffeomorphism F on $\mathbb{R}^N$ as above. For given POD modes W_1 let F_{red} be defined by (5.1). Define the POD subset by*

$$\mathcal{W} := \{W_1 x_1 : x_1 \in B_0^{(\ell)}\} \subset W_1(\mathbb{R}^N) \cap B_0^{(N)} \subset \mathbb{R}^N.$$

If the POD subspace $W_1(\mathbb{R}^N)$ is F-invariant, then every F_{red}-invariant measure $\mu^{(\text{red})}$ extends to a measure $\mu^{(\text{pod})} \in \mathcal{M}^1(\mathbb{R}^N)$ that is F-invariant.

Now assume $\mathcal{W}$ is uniformly attracting, i.e. for all bounded sets $B \subset \mathbb{R}^N$ and $\varepsilon > 0$ there is an index $M = M(B, \varepsilon)$ such that

$$\text{dist}(F^m(B), \mathcal{W}) \leq \varepsilon \quad \text{for all } m \geq M.$$

Then every F-invariant measure $\mu_{\text{inv}} \in \mathcal{M}^1(\mathbb{R}^N)$ is of the form

$$\mu_{\text{inv}} = \left(\mu_\ell \times \delta_0^{N-\ell}\right) \circ W^T$$

with an F_{red}-invariant measure $\mu_\ell \in \mathcal{M}^1(\mathbb{R}^\ell)$.

The second part of Proposition 5.1 is based on the following observation concerning the support of invariant measures (see [CKR08]).

Lemma 5.2. *Let μ_{inv} be an invariant measure of a dynamical system in a Hilbert space H defined by $S : \mathbb{T} \times V \to V$ where V, H define a Gelfand triple $V \subset H \subset V^*$. Let $E \subset V$ be a uniformly absorbing set, i.e. for all $R \geq 0$ there is a time $t_R \in \mathbb{T}$, $t_R > 0$, such that*

$$S(t_R)u_0 \in E \quad \text{for all } u_0 \in V \text{ with } \|u_0\| \leq R.$$

Then,

$$\mu(E) = 1.$$

Proof of Proposition 5.1. First observe that F-invariance of $W_1(\mathbb{R}^N)$ implies $F(W_1 x_1) \in W_1(\mathbb{R}^N)$ and we get

$$W_2 F(W_1 x_1)) = 0 \quad \text{for all } x_1 \in \mathbb{R}^\ell.$$

Thus, for every $x_1 \in \mathbb{R}^\ell$,

$$\begin{aligned} F(W_1 x_1) &= W W^T F(W_1 x_1) = (W_1 W_1^T + W_2 W_2^T) F(W_1 x_1) \\ &= W_1 W_1^T F(W_1 x_1) = W_1 F_{\text{red}}(x_1). \end{aligned} \tag{5.8}$$

It follows that for given F_{red}-invariant measures $\mu^{(\text{red})}$ the extended measure $\mu^{(\text{pod})}$ defined by $\mu^{(\text{pod})} = \left(\mu^{(\text{red})} \times \delta_0^{N-\ell}\right) \circ W^T$ is F-invariant. This can be seen by the following calculation for a Borel set $A \subset \mathbb{R}^N$.

$$\mu^{(\text{pod})}(F^{-1}(A)) = \left(\mu^{(\text{red})} \times \delta_0^{N-\ell}\right)(W^T(F^{-1}(A)))$$

$$\begin{aligned} \overset{\text{(Thm.}}{\underset{\text{A.28)}}{=}} \ & \int_{\mathbb{R}^\ell} \int_{\mathbb{R}^{N-\ell}} \mathbb{1}_{W^T(F^{-1}(A))} \begin{pmatrix} x_1 \\ x_2 \end{pmatrix} d\delta_0^{N-\ell}(x_2) \, d\mu^{(\text{red})}(x_1) \\ = \ & \int_{\mathbb{R}^\ell} \mathbb{1}_{W^T(F^{-1}(A))} \begin{pmatrix} x_1 \\ 0 \end{pmatrix} d\mu^{(\text{red})}(x_1) = \int_{\mathbb{R}^\ell} \mathbb{1}_{F^{-1}(A)}(W_1 x_1)) \, d\mu^{(\text{red})}(x_1) \\ = \ & \int_{\mathbb{R}^\ell} \mathbb{1}_A(F(W_1 x_1)) \, d\mu^{(\text{red})}(x_1) \overset{(5.8)}{=} \int_{\mathbb{R}^\ell} \mathbb{1}_A(W_1 F_{\text{red}}(x_1)) \, d\mu^{(\text{red})}(x_1) \\ = \ & \int_{\mathbb{R}^\ell} \mathbb{1}_{W_1^{-1}(A)}(F_{\text{red}}(x_1)) \, d\mu^{(\text{red})}(x_1) \ = \ \mu^{(\text{red})}(F_{\text{red}}^{-1}(W_1^{-1}(A))) \\ = \ & \mu^{(\text{red})}(W_1^{-1}(A)) \ = \int_{\mathbb{R}^\ell} \mathbb{1}_A(W_1 x_1) \, d\mu^{(\text{red})}(x_1) \\ = \ & \int_{\mathbb{R}^\ell} \int_{\mathbb{R}^{N-\ell}} \mathbb{1}_{W^T(A)} \begin{pmatrix} x_1 \\ x_2 \end{pmatrix} d\delta_0^{N-\ell}(x_2) \, d\mu^{(\text{red})}(x_1) \\ = \ & \left(\mu^{(\text{red})} \times \delta_0^{N-\ell}\right)(W^T(A)) \ = \ \mu^{(\text{pod})}(A). \end{aligned}$$

Here as before in Chapter 1, the matrices W, W_1 are identified with their linear transformations, in particular $W^T(A) = W^{-1}(A)$ and $W_1^{-1}(A)$ denote the preimages of a proper measurable set A.

For the second statement, let $\mu_{\text{inv}} \in \mathcal{M}^1(\mathbb{R}^N)$ be an F-invariant measure. Observe that due to the uniform attraction of $\mathcal{W}$ the neighborhood

$$\mathcal{W}^1 = \{x \in \mathbb{R}^N : d(x, \mathcal{W}) < 1\}$$

is uniformly absorbing, hence, by Lemma 5.2,

$$\mu_{\text{inv}}(\mathcal{W}^1) = 1.$$

The uniform attraction further implies the existence of an $M = M(\mathcal{W}^1, 1)$ such that

$$F^m(\mathcal{W}^1) \subset \mathcal{W}^1 \quad \text{for all } m \geq M,$$

hence, $\{\mathcal{W}_k\}_k$ with $\mathcal{W}_k = F^{kM}(\mathcal{W}^1)$ is a monotone sequence of open sets converging to $\mathcal{W}$. Observe that μ_{inv} is regular, cf. the remark to Definition A.31. Since Lemma 5.2 also applies to $\mathcal{W}_k$ instead of $\mathcal{W}^1$ it follows

$$\mu_{\text{inv}}(\mathcal{W}) = \lim_{k \to \infty} \mu_{\text{inv}}(\mathcal{W}_k) = 1. \tag{5.9}$$

In particular by (5.9), the support of μ_{inv} is a subset of $\mathcal{W}$ and we can represent the invariant measure as

$$\mu_{\text{inv}} = \left(\mu_\ell \times \delta_0^{N-\ell}\right) \circ W^T$$

with a probability measure $\mu_\ell \in \mathcal{M}^1(\mathbb{R}^\ell)$. The following calculation for a Borel set $B \subset \mathbb{R}^N$ shows the F_{red}-invariance of μ_ℓ:

$$
\begin{aligned}
\mu_\ell(F_{\text{red}}^{-1}(B)) &= \int_{\mathbb{R}^\ell} \mathbb{1}_{F_{\text{red}}^{-1}(B)}(x_1)\, d\mu_\ell(x_1) \\
&= \int_{\mathbb{R}^\ell} \mathbb{1}_B(W_1^T F(W_1 x_1))\, d\mu_\ell(x_1) \\
&= \int_{\mathbb{R}^\ell} \mathbb{1}_{B \times \mathbb{R}^{N-\ell}}\left(W^T F\left(W \begin{pmatrix} x_1 \\ 0 \end{pmatrix}\right)\right) d\mu_\ell \times \delta_0^{N-\ell}\begin{pmatrix} x_1 \\ x_2 \end{pmatrix} \\
&= \int_{\mathbb{R}^N} \mathbb{1}_{W(B \times \mathbb{R}^{N-\ell})}\left(F\left(W \begin{pmatrix} x_1 \\ x_2 \end{pmatrix}\right)\right) d\delta_0^{N-\ell}(x_2), d\mu_\ell(x_1) \\
&= \left(\mu_\ell \times \delta_0^{N-\ell}\right)\left(W^T\left(F^{-1}\left(W\left(B \times \mathbb{R}^{N-\ell}\right)\right)\right)\right) \\
&= \mu_{\text{inv}}\left(F^{-1}\left(W\left(B \times \mathbb{R}^{N-\ell}\right)\right)\right) \quad = \mu_{\text{inv}}\left(W\left(B \times \mathbb{R}^{N-\ell}\right)\right) \\
&= \mu_\ell \times \delta_0^{N-\ell}\left(B \times \mathbb{R}^{N-\ell}\right) \qquad = \mu_\ell(B).
\end{aligned}
$$

$\square$

Remark. *In general, the whole dynamics of the original systems will not be embedded in the low-dimensional POD space as assumed in Proposition 5.1. Hence, the question arises of how to generalize the results. One approach is as follows. For $\varepsilon > 0$ arbitrarily small, assume there is a POD space $\mathcal{W}$ approximating the global attractor $\mathcal{A}$ of the system in the sense that*

$$d_{NH}(\mathcal{A}, \mathcal{W}) < \varepsilon$$

for the unsymmetric Hausdorff distance d_{NH}. If the system is structurally stable, one would expect a similar behavior on $\mathcal{W}$. Indeed, numerical experiments suggest such a behavior of the POD-based model reduction. A rigorous proof of this observation for structurally stable systems in general is beyond our scope. Instead, we analyze the case where F has a proper block structure.

5.2.2 Relations between the transfer operators

Now we assume that we can decompose F in a nice way. We show that this decomposition passes on to the Frobenius-Perron operator.

Proposition 5.3. *Let there exist a decomposition of $F : \mathbb{R}^N \to \mathbb{R}^N$ into the direct sum*

$$F(v + w) = F(v) + F(w), \quad v \in \mathcal{W}, w \in \mathcal{W}^\perp \tag{5.10}$$
$$\text{with} \qquad F(\mathcal{W}) = \mathcal{W}, \; F(\mathcal{W}^\perp) = \mathcal{W}^\perp.$$

Let the Frobenius-Perron operator for F, F_{red} and F_{red_2} with $F_{\text{red}_2}(x_2) = W_2^T F(W_2 x_2)$, $x_2 \in \mathbb{R}^{N-\ell}$, be given by P, $P^{(\text{red})}$ and $P^{(\text{red}_2)}$, respectively. Then for every product measure $\mu = (\mu_1 \times \mu_2) \circ W^T \in \mathcal{M}^1(\mathbb{R}^N)$ with $\mu_1 \in \mathcal{M}^1(\mathbb{R}^\ell)$ we get the following relation of the transfer operators:

$$(P\mu) \circ W = P^{(\text{red})}\mu_1 \times P^{(\text{red}_2)}\mu_2. \tag{5.11}$$

Proof. Define the orthogonal projection P_W onto W according to Definition A.15 by $P_W x = W_1 W_1^T x$. Then (5.10) implies

$$
\begin{aligned}
F(x) &= F(P_W x + (I - P_W)x) = F(P_W x) + F((I - P_W)x) \\
&= P_W F(P_W x) + (I - P_W)F((I - P_W)x). \\
&= W_1 F_{\mathrm{red}}(W_1^T x) + W_2 F_{\mathrm{red}_2}(W_2^T x).
\end{aligned}
\tag{5.12}
$$

Now let $\mu = (\mu_1 \times \mu_2) \circ W^T \in \mathcal{M}^1(\mathbb{R}^N)$ be given with $\mu_1 \in \mathcal{M}^1(\mathbb{R}^\ell)$. For cylinder sets $A_1 \times A_2 \in \mathcal{B}(\mathbb{R}^N)$, $A_1 \in \mathcal{B}(\mathbb{R}^\ell)$, $A_2 \in \mathcal{B}(\mathbb{R}^{N-\ell})$, we have

$$
\begin{aligned}
P\mu(W(A_1 \times A_2)) &= \mu(F^{-1}(W(A_1 \times A_2))) \\
&= (\mu_1 \times \mu_2)(W^T(F^{-1}(W(A_1 \times A_2)))) \\
&= \int_{\mathbb{R}^N} \mathbb{1}_{W(A_1 \times A_2)}(F(Wx)) \, d(\mu_1 \times \mu_2)(x) \\
&= \int_{\mathbb{R}^\ell} \int_{\mathbb{R}^{N-\ell}} \mathbb{1}_{W(A_1 \times A_2)}\big(F(W_1 x_1 + W_2 x_2)\big) \, d\mu_2(x_2) \, d\mu_1(x_1) \\
&\overset{(5.12)}{=} \int_{\mathbb{R}^\ell} \int_{\mathbb{R}^{N-\ell}} \mathbb{1}_{A_1 \times A_2} W^T\big(W_1 F_{\mathrm{red}}(x_1) + W_2 F_{\mathrm{red}_2}(x_2)\big) d\mu_2(x_2) \, d\mu_1(x_1) \\
&= \int_{\mathbb{R}^\ell} \mathbb{1}_{A_1}(F_{\mathrm{red}}(x_1)) \, d\mu_1(x_1) \int_{\mathbb{R}^{N-\ell}} \mathbb{1}_{A_2}(F_{\mathrm{red}}(x_2)) \, d\mu_2(x_2) \\
&= \int_{\mathbb{R}^\ell} \mathbb{1}_{A_1}(x_1) \, d(P^{(\mathrm{red})}\mu_1)(x_1) \int_{\mathbb{R}^{N-\ell}} \mathbb{1}_{A_2}(x_2) \, d(P^{(\mathrm{red}_2)}\mu_2)(x_2) \\
&= \int_{\mathbb{R}^N} \mathbb{1}_{A_1 \times A_2}(x) \, d\big(P^{(\mathrm{red})}\mu_1 \times P^{(\mathrm{red}_2)}\mu_2\big)(x) \\
&= \big(P^{(\mathrm{red})}\mu_1 \times P^{(\mathrm{red}_2)}\mu_2\big)(A_1 \times A_2).
\end{aligned}
$$

By a limit process the same equality holds for all $A \in \mathcal{B}(\mathbb{R}^N)$, thus, (5.11) holds. $\qquad\square$

Proposition 5.3 suggests that a decomposition might work for continuous measures in order to get error bounds of the POD method. In general the dynamics are obviously not as well separated as given by assumption (5.10). In most cases, the projections of an invariant measure to the according subspaces are no longer probability measures. Thus, if one pursues this approach one has to work with signed measures to set up a block structure for the involved measures in the non-diagonal case and use a diagonalization argument. We do not make further steps in this direction since such a block structure is not natural in the space of probability measures. Instead, we focus on the discrete case where the transfer operators are given by (finite-dimensional) Perron-Frobenius matrices and the block structures occur more naturally.

5.3   Error bounds for the discrete measures

In this section, we derive error bounds for discrete measures in the POD subspace in terms of discrete measures of the AIM algorithm. As mentioned above, a promising approach is given by the decomposition of the Perron-Frobenius matrix P_k of the original system according to the POD subspace and comparing one diagonal block with the Perron-Frobenius matrix $P_k^{(\mathrm{red})}$ of the reduced-order algorithm. Since μ_k and $\mu_k^{(\mathrm{red})}$ are given by fixed points of the corresponding Perron-Frobenius matrices P_k and $P_k^{(\text{red})}$, respectively, distances between these measures are given by distances between the Perron roots of the matrices. Therefore we introduce some theory about perturbation analysis concerning the spectrum of stochastic matrices in the following.

5.3.1   Perturbations of Perron roots

We are interested in perturbation results for the fixed points of Perron-Frobenius matrices, i.e. eigenvectors associated with the eigenvalue 1, sometimes just called Perron roots.

For an irreducible Perron-Frobenius matrix P the eigenvalue 1 is simple, hence the standard perturbation theory for eigenvectors of complex matrices applies. Standard results (see [GvL96], [Ste73]) for arbitrary matrices are of the following type: Consider $A, E \in \mathbb{C}^{n,n}$, $\tilde{A} = A + E$ and $\lambda \in \mathbb{C}$ a simple eigenvalue of A with spectral gap

$$\text{gap}(\lambda, A) := \min\{|\lambda - \mu| : \mu \in \sigma(A) \setminus \{\lambda\}\},$$

and eigenvector $x \in \mathbb{C}^n$. Then there is an eigenpair (λ', x') with λ' close to λ and $\|x'\|_2 = 1$ such that

$$\|x - x'\|_2 \leq \frac{\varepsilon}{\text{gap}(\lambda, A)}.$$

Here, $\varepsilon > 0$ grows with the so-called *condition* $\kappa(A) = \|A\|_2\|A^{-1}\|_2$ *of the matrix* A.

However, since Perron-Frobenius matrices provide some special structure, being nonnegative and stochastic, we expect better approximation results for the Perron root than given by the general theory. Indeed there are plenty of results concerning the Perron roots of perturbed nonnegative matrices.

For arbitrary (possibly reducible) nonnegative matrices the introduction of some generalized inverses is useful. See [CM79], [BIG03] for a broad theory of generalized inverses.

Definition 5.4. *For* $A \in \mathbb{C}^{n,n}$ *we define the* index of A at $\lambda \in \mathbb{C}$ *by*

$$\text{ind}_\lambda(A) := \min\{k \in \mathbb{N} : N((A - \lambda I_n)^k) = N((A - \lambda I_n)^{k+1})\}.$$

In particular $\text{ind}_0(A)$ is just called the index of A.

For a matrix A with index $\ell = \text{ind}_0(A)$ the *Drazin inverse* A^D is defined by the properties

$$A^D A = AA^D, \quad A^D A A^D = A^D, \quad A^\ell = A^{\ell+1} A^D.$$

For $\ell = 1$ the Drazin inverse is called *group inverse* and denoted by $A^\#$. The defining properties can be written as

$$A^\# A = AA^\#, \quad A^\# A A^\# = A^\#, \quad AA^\# A = A.$$

The Drazin inverse is well defined for any matrix $A \in \mathbb{C}^{n,n}$. Observe that in particular $A^D = A^{-1}$ for $\ell = 0$.

Lemma 5.5 ([CM79]). *For* $A \in \mathbb{C}^{n,n}$ *with* $\text{ind}_0(A) \geq 1$ *denote the eigenvalues by* $0 = \lambda_0, \lambda_1, \ldots, \lambda_k$ *with corresponding generalized eigenspaces* $E(\lambda_i)$, $i = 1, \ldots, k$, *such that* $\bigoplus_{i=0}^{k} E(\lambda_i) = \mathbb{C}^n$. *Then* $AA^D = A^D A$ *is a linear projection with*

$$N(AA^D) = E(0), \quad R(AA^D) = \bigoplus_{i=1}^{k} E(\lambda_i).$$

Lemma 5.5 immediately defines a proper projection onto the generalized eigenspace corresponding to the Perron root $r = \rho(P)$ of a nonnegative matrix P by

$$Z_P = I - (rI - P)(rI - P)^D. \tag{5.13}$$

In [HNR90] this approach is used to derive nonnegative bases of the generalized eigenspace of the Perron root of P. We will come back to this approach later on in chapter 6 when we face the problem that the Perron-Frobenius matrix might be reducible in applications.

For a Perron-Frobenius matrix $P \in \mathbb{R}^{n,n}$, i.e. P stochastic and nonnegative, it is well-known (see [CM79]), that $\mathrm{ind}_1(P) = 1$, so that (5.13) can be expressed in terms of the group inverse,

$$Z_P = I - (I - P)(I - P)^{\#}. \tag{5.14}$$

One concept to derive perturbation estimates for reducible Perron-Frobenius matrices is given by using this representation for matrices $P, \tilde{P} = P + E$, since (5.14) implies

$$\|Z_P - Z_{\tilde{P}}\|_{\infty} = \|(I - P)(I - P)^{\#} - (I - \tilde{P})(I - \tilde{P})^{\#}\|. \tag{5.15}$$

Indeed group inverses behave well under perturbation (while Drazin inverses in general do not), such that bounds for the right hand side of (5.15) exist. For example, the following proposition in [LW01] gives a relative bound for perturbations of projections of the type $AA^{\#}$.

Proposition 5.6. *Let $A, E \in \mathbb{C}^{n,n}$, $\tilde{A} = A + E$, with $\mathrm{ind}_0(A) \leq 1$ and $\mathrm{rank}(\tilde{A}) = \mathrm{rank}(A)$. If*

$$\|A^{\#}\|\|E\| < \frac{1}{1 + \mathrm{ind}_0(A)\|AA^{\#}\|},$$

then

$$\frac{\|\tilde{A}\tilde{A}^{\#} - AA^{\#}\|}{\|AA^{\#}\|} \leq \frac{\|A^{\#}\|\|E\|(1 + \mathrm{ind}_0(A)\|A^{\#}\|\|E\|\|AA^{\#}\|)}{(1 - \|A^{\#}\|\|E\|)^2 - (\mathrm{ind}_0(A)\|A^{\#}\|\|E\|\|AA^{\#}\|)^2}.$$

For irreducible Perron-Frobenius matrices, Funderlic and Meyer showed a more direct perturbation result for the Perron roots in [Mey80], [FM86]:

Proposition 5.7. *Let $P, Q = P + E \in \mathbb{R}^{n,n}$ be irreducible stochastic matrices. Then the fixed points $\bar{u}, \bar{v} \in \mathbb{R}^n$, $\bar{u}, \bar{v} \geq 0$, of P and Q satisfy*

$$|\bar{u} - \bar{v}|_i \leq \|E\|_{\infty} \max_{j=1,\ldots,n} |a_{ij}^{\#}| \tag{5.16}$$

where $A^{\#} = (a_{ij}^{\#})_{ij}$ is the group inverse of $A = I - P$.

In particular, (5.16) implies $\|\bar{u} - \bar{v}\|_{\infty} \leq \bar{\kappa}(P)\|E\|_{\infty}$ where

$$\bar{\kappa}(P) = \max_{i,j=1,\ldots,n} |a_{ij}^{\#}|. \tag{5.17}$$

This motivates to call $\bar{\kappa}(P)$ the *condition number* of the Perron-Frobenius matrix P in terms of the Perron root. Note that Meyer shows in [Mey80] that this condition number might be small for irreducible Perron-Frobenius matrices P in cases where $\kappa(P) = \|P\|\|P^{-1}\|$ is arbitrarily large. Thus Proposition 5.7 provides significantly better results than the general eigenvector perturbation theory in the case of irreducible matrices. To derive the condition number there is a suitable result in [Mey94] relating the condition number of P to the so-called character of a stochastic matrix.

Proposition 5.8. *Let* $P \in \mathbb{R}^{n,n}$ *be an irreducible stochastic matrix with eigenvalues* $1 = |\lambda_1| \geq |\lambda_2| \geq \ldots \geq |\lambda_n| \geq 0$ *counted with multiplicity. For* $A = (a_{ij})_{ij} = I - P$ *and* $1 \leq i,j \leq n$, $i \neq j$ *define the number*

$$\delta_{ij}(A) := \prod_{k \neq i,j}^{n} a_{kk}$$

and set $\delta := \max\{\delta_{ij}(A) : 1 \leq i,j \leq n, i \neq j.\}$. *Then the following bounds hold for* $\bar{\kappa}(P)$:

$$\frac{1}{n(1-\lambda_2)} \leq \bar{\kappa}(P) < \frac{2\delta(n-1)}{\chi(P)} \leq \frac{2(n-1)}{\chi(P)}. \tag{5.18}$$

Here, $\chi(P)$ *is the* character *of* P *defined by*

$$\chi(P) := (1-\lambda_2) \cdot \ldots \cdot (1-\lambda_n) \in (0,n]. \tag{5.19}$$

We will use this approach at the end of the section to provide a perturbation result for the reduced-order discrete measure $\mu_k^{(\mathrm{red})}$ in terms of the discrete measure μ_k of the original system derived by the AIM algorithm.

5.3.2   Perron-Frobenius matrices of reduced systems

For comparing discrete measures $\mu_k^{(\mathrm{pod})}$ in the POD subspace and discrete measures μ_k arising in the AIM algorithm, we decompose the Perron-Frobenius matrix P_k where $P_k \mu_k = \mu_k$.

Therefore we start with some notations and establish a first general assumption. Consider the box collection $\mathcal{B} = \mathcal{B}_k^{(N)}$ in recursion step $k \in \mathbb{N}$ of the AIM algorithm consisting of $K = K_N$ boxes. We assume that no boxes are eliminated in the refinement process, i.e. $\mathcal{B}$ covers $B_0^{(N)}$. By a rearrangement of the boxes we get a distribution $\mathcal{B} = \mathcal{B}_{\mathrm{I}} \cup \mathcal{B}_{\mathrm{II}}$ with

$$\mathcal{B}_{\mathrm{I}} = \{B_1, \ldots, B_{K_1}\} \quad = \{B \in \mathcal{B} : B \cap \mathcal{W} \neq \emptyset\}$$
$$\mathcal{B}_{\mathrm{II}} = \{B_{K_1+1}, \ldots, B_K\} = \{B \in \mathcal{B} : B \cap \mathcal{W} = \emptyset\}$$

where $\mathcal{W} = \{W_1 x_1 : x_1 \in B_0^{(\ell)}\}$ denotes the POD subset as in Proposition 5.1. According to the distribution of the boxes we get a natural block structure of the Perron-Frobenius matrix $P_k \in \mathbb{R}^{K,K}$ by

$$P_k =: \begin{pmatrix} P_{\mathrm{I,I}} & P_{\mathrm{I,II}} \\ P_{\mathrm{II,I}} & P_{\mathrm{II,II}} \end{pmatrix} \tag{5.20}$$

with square matrices $P_{\mathrm{I}} \in \mathbb{R}^{K_1,K_1}$, $P_{\mathrm{II,II}} \in \mathbb{R}^{K-K_1,K-K_1}$ and $P_{\mathrm{I,II}} \in \mathbb{R}^{K_1,K-K_1}$, $P_{\mathrm{II,I}} \in \mathbb{R}^{K-K_1,K_1}$. We need the following assumptions on the discrete dynamical system defined by F.

Assumption C$_1$. Assume:

- The subset $\mathcal{W}$ uniformly attracts $B_0^{(N)}$.

- The reduced-space starting box $B_0^{(\ell)}$ is embedded in the interior of $B_0^{(N)}$ in the sense that

$$\mathcal{W} = W_1(B_0^{(\ell)}) \subset \mathrm{int}(B_0^{(N)}).$$

- The box collection $\mathcal{B}_I$ is positive invariant under F, $F(\mathcal{B}_I) \subset \mathcal{B}_I$.

Proposition 5.9. *We assume $\mathbf{C_1}$ with notions as above. Then every fixed point $\bar{u}$ of P_k decomposes into*

$$\bar{u} = \begin{pmatrix} \bar{u}_I \\ 0 \end{pmatrix} \tag{5.21}$$

where $\bar{u}_I$ is a fixed point of $P_{I,I}$. In particular, the support of the discrete measure μ_k corresponding to P_k is located in the box covering $\mathcal{B}_I$ of $\mathcal{W}$.

Proof. Since by assumption $\mathcal{B}_I$ is positive invariant under F, we obtain

$$F^{-1}(B_i) \subset \mathcal{B}_I^c \quad \text{for all } B_i \in \mathcal{B}_{II}.$$

Then by definition of $P_k = (p_{ij})_{ij}$, we get:

$$p_{ij} = \frac{\lambda_N(F^{-1}(B_i) \cap B_j)}{\lambda_N(B_j)} = 0 \quad \text{for } 1 \le j \le K_1 < i \le K.$$

Hence, the lower left block of P_k vanishes:

$$P_k = \begin{pmatrix} P_{I,I} & P_{I,II} \\ 0 & P_{II,II} \end{pmatrix}. \tag{5.22}$$

Observe that this implies that the powers of P_k can be decomposed into

$$P_k^n = \begin{pmatrix} * & * \\ 0 & P_{II,II}^n \end{pmatrix}$$

for every $n \in \mathbb{N}$. We will use the equality (A.2) of the Perron-Frobenius Theorem in the version of column sums. Therefore we define the column sum $s_j^{(n)}$, $K_1 < j \le K$ of $P_{II,II}^n = (p_{ij}^{(n)})_{k_1 < i,j \le K}$, $n \in \mathbb{N}$ by

$$s_j^{(n)} = \sum_{i=K_1+1}^{K} p_{ij}^{(n)}.$$

Since P_k^m is column stochastic, an upper bound for $s_j^{(n)}$ is given by 1:

$$0 \le s_j^{(n)} \le 1 \quad \text{for all } K_1 < j \le K, \ n \in \mathbb{N}.$$

Assume now the following implication for every $K_1 < j \le K$, $n \in \mathbb{N}$

$$s_j^{(n)} = 1 \overset{!}{\Longrightarrow} B_j \subset F^{-n}(\mathcal{B}_{II}) \cup N_\lambda \tag{5.23}$$

for a Lebesgue-null set $N_\lambda \subset \mathbb{R}^N$.

Since the boxes are compact sets by definition, it holds that

$$\partial \mathcal{B}_I \subset \mathcal{B}_{II}$$

in the interior $\text{int}(B_0^{(N)})$ of the starting box. Hence, the defining property $\mathcal{W} \cap \mathcal{B}_{II} = \emptyset$ of the second box collection implies

$$\mathcal{W} \cap \partial \mathcal{B}_I = \emptyset.$$

Thus the compact set $\mathcal{W}$ is located in the interior of the compact set $\mathcal{B}_I$ and we can find an $\epsilon > 0$ with

$$\mathcal{W}^\varepsilon = \{v \in \mathbb{R}^N : d(v, \mathcal{W}) < \varepsilon\} \subset \mathcal{B}_I.$$

Further by assumption $\mathbf{C_1}$, the starting box $B_0^{(N)}$ is uniformly attracted by $\mathcal{W}$. In particular, there exists an $\hat{M} = M(B_0^{(N)}, \varepsilon)$ with $F^M(\mathcal{B}_{II}) \subset F^M(B_0^{(N)}) \subset \mathcal{B}_I$. Since F is diffeomorph, it follows that

$$\mathcal{B}_{II} \subset F^{-M}(\mathcal{B}_I) \quad \Longrightarrow \quad \mathcal{B}_{II} \cap F^{-M}(\mathcal{B}_{II}) = N_\lambda$$

for a Lebesgue-null set $N_\lambda \subset \mathbb{R}^N$. Provided (5.23) holds, this implies that all column sums of $P_{II,II}^M$ are smaller than 1. Hence by the Perron-Frobenius Theorem it follows for the nonnegative matrix $P_{II,II}^M$ that

$$\rho(P_{II,II}^M) \leq \max_{K_1 < j \leq K} s_j^M < 1.$$

In particular, only the trivial fixed point of $P_{II,II}$ exists. But for a fixed point $\bar{u} = \begin{pmatrix} \bar{u}_I \\ \bar{u}_{II} \end{pmatrix}$ of P_k, (5.22) implies

$$P_{II,II}\bar{u}_{II} = \bar{u}_{II}.$$

Hence, every fixed point of P_k is of the type (5.21).

At the end of the proof we show (5.23) by induction over n. The initial case $n = 0$ is trivial since $B_j \subset \mathcal{B}_{II}$ by definition.

Assume (5.23) holds for arbitrary $n \in \mathbb{N}$ and $s_j^{(n+1)} = 1$ for some j, $K_1 < j \leq K$. Then by matrix multiplication, we get

$$
\begin{aligned}
1 = s_j^{(n+1)} &= \sum_{i=K_1+1}^{K} p_{ij}^{(n+1)} && = \sum_{i=K_1+1}^{K} \sum_{\ell=K_1+1}^{K} p_{i\ell}^{(n)} p_{\ell j} \\
&= \sum_{\ell=K_1+1}^{K} p_{\ell j} \sum_{i=K_1+1}^{K} p_{i\ell}^{(n)} = \sum_{\ell=K_1+1}^{K} p_{\ell j} s_\ell^{(n)}.
\end{aligned}
\tag{5.24}
$$

Since $\sum_{\ell=K_1+1}^{K} p_{\ell j} = s_j^{(1)} \leq 1$, $p_{\ell j} \geq 0$ and $s_\ell^{(n)} \leq 1$, the last term in (5.24) is a convex sum of numbers in $[0, 1]$. This convex sum equals one iff

$$s_j^{(1)} = 1 \quad \text{and} \tag{5.25}$$

$$p_{\ell j} = 0 \quad \text{for all } K_1 < \ell \leq K \text{ with } s_\ell^{(n)} < 1. \tag{5.26}$$

Property (5.25) implies $B_j \subset F^{-1}(\mathcal{B}_{II}) \cup N_\lambda$. By definition of $p_{\ell j}$, (5.26) implies that $B_j \cap F^{-1}(B_\ell)$ is a Lebesgue-null set for every $K_1 < \ell \leq K$ with $s_\ell^{(n)} < 1$. All in all we get

$$B_j \overset{(5.25)}{\subset} \bigcup_{\ell=K_1+1}^{K} F^{-1}(B_\ell) \cup N_\lambda \overset{(5.26)}{\subset} \bigcup_{\substack{\ell=K_1+1 \\ s_\ell^{(n)}=1}}^{K} F^{-1}(B_\ell) \cup N_\lambda$$

$$\text{(induction hypothesis)} \quad = \bigcup_{\substack{\ell=k_1+1 \\ B_\ell \subset F^{-n}(\mathcal{B}_{II}) \cup N_\lambda}}^{K} F^{-1}(B_\ell) \cup N_\lambda \subset F^{-(n+1)}(\mathcal{B}_{II}) \cup N_\lambda$$

where N_λ is a generic Lebesgue-null set. $\square$

Proposition 5.9 allows us to focus our considerations on the box collection $\mathcal{B}_I$ covering the POD subset $\mathcal{W}$. We set up some more regularity assumptions by which we are able to show that the diagonal block $P_{I,I}$ of P_k can be written as a perturbation of the Perron-Frobenius matrix $P_k^{(\text{red})}$ of the reduced system.

Assumption C_2. Assume:

- The POD subspace $W_1(\mathbb{R}^N)$ is invariant and paraxial, i.e.

$$W_1 = \text{col}(e_1, \ldots, e_\ell),$$

  where e_i are the standard basis vectors in $\mathbb{R}^N$.

- The cardinalities of $\mathcal{B}_I$ and $\mathcal{B}_k^{(\ell)}$ are the same, $K = K_N = K_\ell$. Moreover the box collections are of the following type:

  1. $\mathcal{B}_k^{(\ell)} = \{B_1^{(\ell)}, \ldots, B_K^{(\ell)}\}$ covers $\mathcal{W}$ with ℓ hypercubes of the same radius:

  $$B_i^{(\ell)} = B(c_i, r\mathbb{1}).$$

  2. $\mathcal{B}_I = \{B_1^{(N)}, \ldots, B_K^{(N)}\}$ consists of blown-up versions of the boxes in $\mathcal{B}_k^{(\ell)}$:

  $$B_i^{(N)} = B_i^{(\ell)} \times [-\varepsilon, \varepsilon]^{N-\ell} = B\left(\begin{pmatrix} c_i \\ 0 \end{pmatrix}, \begin{pmatrix} r\mathbb{1} \\ \varepsilon\mathbb{1} \end{pmatrix}\right), \quad i = 1, \ldots, K. \qquad (5.27)$$

Remark. *By the special shape of the POD subspace $W_1(\mathbb{R}^N)$ we get a simplified version of the reduced system function F_{red} since $W_1 = \text{col}(e_1, \ldots, e_\ell)$ implies*

$$\begin{pmatrix} F_{\text{red}}(u_1) \\ 0 \end{pmatrix} = \begin{pmatrix} W_1^T F(W_1 u_1) \\ 0 \end{pmatrix} = \begin{pmatrix} W_1^T F\left(\begin{pmatrix} u_1 \\ 0 \end{pmatrix}\right) \\ 0 \end{pmatrix} = F\begin{pmatrix} u_1 \\ 0 \end{pmatrix}, \quad u_1 \in \mathbb{R}^\ell. \qquad (5.28)$$

The last equality follows from the F-invariance of $W_1(\mathbb{R}^N)$. This implies in particular that F_{red} is a bijection.

For the next proposition we will use the following notation for the closed ε-neighborhood of a compact set B, cf. (4.2) for the open neighborhood on a general metric space.

Definition 5.10. Let B be a compact set in $\mathbb{R}^n$, $n \in \mathbb{N}$, and $\varepsilon > 0$. We denote the closed ε-neighborhood of B with respect to the norm $\|\cdot\|_\infty$ by $B^{+\varepsilon}$. More precisely,

$$B^{+\varepsilon} := \overline{B^\varepsilon} = \{x \in \mathbb{R}^n : \inf_{y \in B} \|x - y\|_\infty \le \varepsilon\}. \qquad (5.29)$$

We will also use a shrunk version of (5.29) which we denote by $B^{-\varepsilon}$:

$$B^{-\varepsilon} := ((\overline{B^c})^{+\varepsilon})^c = \{x \in \mathbb{R}^n : \inf_{y \in B^c} \|x - y\|_\infty > \varepsilon\}.$$

In our context, we use this definition for $n \in \{N, \ell\}$. Under the assumptions $\mathbf{C}_1$ and $\mathbf{C}_2$ we can formulate a perturbation result for the K-dimensional Perron-Frobenius matrices $P_{I,I}$ and $P_k^{(\text{red})}$.

Theorem 5.11. *Let $P_{\mathrm{I,I}} = (p_{ij})_{ij}$ and $P_k^{(\mathrm{red})} = (p_{ij}^{(\mathrm{red})})_{ij}$ be the Perron-Frobenius matrices as defined above for the dynamical systems given by F and F_{red}. Denote Lipschitz constants M_1, M_2 of F, F^{-1} on the closed δ-neighborhood of $\mathcal{W}$, $\delta > 0$, by*

$$M_1 := \max\left(\sqrt{2}, \max_{x \in \mathcal{W}^{+\delta}} \|DF(x)\|_\infty\right), \quad M_2 := \max\left(\sqrt{2}, \max_{x \in \mathcal{W}^{+\delta}} \|DF^{-1}(x)\|_\infty\right). \quad (5.30)$$

For given

$$0 < \varepsilon \le \varepsilon_0 := \min\left(\alpha \frac{2r}{3M_1}, \frac{2\delta}{M_1 M_2}\right)$$

with an arbitrary $\alpha > 0$, assume the box collections $\mathcal{B}_k^{(\ell)}$ and $\mathcal{B}_{\mathrm{I}} = \mathcal{B}_{\mathrm{I}}(\varepsilon)$ satisfy assumptions $\mathbf{C_1}$ and $\mathbf{C_2}$. Then the following error bound holds for the entries of $P_{\mathrm{I}} = P_{\mathrm{I}}(\varepsilon)$ and $P_k^{(\mathrm{red})}$:

$$|p_{ij} - p_{ij}^{(\mathrm{red})}| \le \varepsilon \frac{\ell M_1}{r} (1+\alpha)^{\ell-1} M_2^\ell. \quad (5.31)$$

In particular, for the induced matrix norm $\|\cdot\|_\infty$, we get the estimate

$$\|P_{\mathrm{I,I}} - P_k^{(\mathrm{red})}\|_\infty \le \varepsilon \frac{K\ell M_1}{r} (1+\alpha)^{\ell-1} M_2^\ell. \quad (5.32)$$

The idea of the proof is to treat a box $B_i^{(N)}$ as part of the closed ε-neighborhood of $B_i^{(\ell)}$ and control the error of the preimage. We first need a lemma estimating preimages of $F_{\mathrm{red}} : \mathbb{R}^\ell \to \mathbb{R}^\ell$.

Lemma 5.12. *Let the assumptions of Theorem 5.11 be satisfied. Then for the reduced system function F_{red}, the following implications hold:*

(i) *For a compact set B in $\mathcal{W}$ and $0 < \gamma \le \delta$ with $\delta > 0$ as in Theorem 5.11, we get the embeddings*

$$F_{\mathrm{red}}^{-1}(B^{+\frac{\gamma}{M_2}}) \subset (F_{\mathrm{red}}^{-1}(B))^{+\gamma} \subset F_{\mathrm{red}}^{-1}(B^{+\gamma M_1}). \quad (5.33)$$

(ii) *For hypercubes $B_r = B(c, r\mathbb{1})$ and $B_{r+d} = B(c, (r+d)\mathbb{1})$, where $\frac{1}{2}dM_2 \le \delta$, we get the following estimate*

$$\lambda_\ell(F_{\mathrm{red}}^{-1}(B_{r+d} \backslash B_r)) \in \left[M_1^{-\ell}\left\{(2r+d)^\ell - (2r-d)^\ell\right\}, M_2^\ell\left\{(2r+3d)^\ell - (2r+d)^\ell\right\}\right]. \quad (5.34)$$

Proof. (i) The embedding (5.33) is illustrated in Figure 5.1. Observe the following implications for elements in the POD subspace $W_1(\mathbb{R}^N)$:

Given $v_1 \in F_{\mathrm{red}}^{-1}(B^{+\frac{\gamma}{M_2}})$, then by definition there exists a $u_1 \in F_{\mathrm{red}}^{-1}(B)$ with

$$\|F_{\mathrm{red}}(v_1) - F_{\mathrm{red}}(u_1)\|_\infty \le \frac{\gamma}{M_2}.$$

By (5.28) and the mean value theorem we have

$$\left\|\begin{pmatrix} v_1 \\ 0 \end{pmatrix} - \begin{pmatrix} u_1 \\ 0 \end{pmatrix}\right\|_\infty = \left\|F^{-1}\left(F\begin{pmatrix} v_1 \\ 0 \end{pmatrix}\right) - F^{-1}\left(F\begin{pmatrix} u_1 \\ 0 \end{pmatrix}\right)\right\|_\infty$$

$$\stackrel{(5.28)}{=} \left\|F^{-1}\begin{pmatrix} F_{\mathrm{red}}(v_1) \\ 0 \end{pmatrix} - F^{-1}\begin{pmatrix} F_{\mathrm{red}}(u_1) \\ 0 \end{pmatrix}\right\|_\infty$$

$$\stackrel{(5.30)}{\le} M_2 \|F_{\mathrm{red}}(v_1) - F_{\mathrm{red}}(u_1)\|_\infty \le \gamma.$$

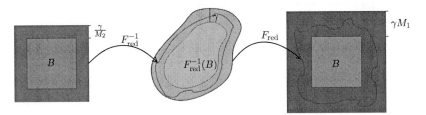

Figure 5.1: Illustration of the embedding (5.33).

Thus $v_1 \in F_{\text{red}}^{-1}(B)^{+\gamma}$.

For arbitrary $v_1 \in F_{\text{red}}^{-1}(B)^{+\gamma}$, there is a $u_1 \in F_{\text{red}}^{-1}(B)$ with $\|v_1 - u_1\|_\infty \leq \gamma$. It follows by

$$\left\| \binom{F_{\text{red}}(v_1)}{0} - \binom{F_{\text{red}}(u_1)}{0} \right\|_\infty \overset{(5.28)}{=} \left\| F\binom{v_1}{0} - F\binom{u_1}{0} \right\|_\infty$$
$$\overset{(5.30)}{\leq} M_1 \left\| \binom{v_1}{0} - \binom{u_1}{0} \right\|_\infty \leq \gamma M_1$$

that $v_1 \in F_{\text{red}}^{-1}(B^{+M_1\gamma})$.

(ii) Observe that the set $H_r := B_{r+d} \setminus B_r$ can be covered by hypercubes $C_i^{(d,r)}$ of the same side length d,

$$C_i^{(d,r)} = B\left(c_i, \frac{d}{2}\right), \quad i = 1, \ldots, N(d,r).$$

We count the number $N(d,r)$ of boxes needed to cover H_r in the following.

As usual for triangulizations, we call the covering $\{C_i\}_{1 \leq i \leq N}$ *regular* iff for every $i, j = 1, \ldots, N$, $i \neq j$ it holds: C_i and C_j either are disjoint or share a common edge or vertex.

If d divides $2r$, it is easy to see, that a regular box covering exists with cardinality

$$N(d,r) = \frac{\lambda_\ell(H_r)}{\lambda_\ell(C_1^{(d,r)})} = \frac{(2r + 2d)^\ell - (2r)^\ell}{(2d)^\ell}.$$

In the general case, no regular box covering exists with hypercubes of the same side length d. Nevertheless, since $r \mapsto \lambda(H_r)$ is monotone, it follows that

$$\sum_{i=1}^{N(d,\underline{r})} \lambda_\ell(C_i^{(d,\underline{r})}) = \lambda(H_{\underline{r}}) \leq \lambda(H_r) \leq \lambda(H_{\overline{r}}) = \sum_{i=1}^{N(d,\overline{r})} \lambda_\ell(C_i^{(d,\overline{r})})$$

$$\text{where} \qquad \underline{r} = \left\lfloor \frac{2r}{d} \right\rfloor \frac{d}{2}, \quad \overline{r} = \left\lceil \frac{2r}{d} \right\rceil \frac{d}{2},$$

where $\{C_i^{(d,\underline{r})}\}_i$ and $\{C_i^{(d,\overline{r})}\}_i$ are regular coverings of $H_{\underline{r}}$ and $H_{\overline{r}}$, respectively. By shifting the boxes of the covering, it follows that a subset of H_r is covered regularly

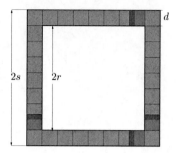

Figure 5.2: Idea of the proof for (5.34). The hull can be approximated from above and below by a quantified number of small n-squares of side length d.

by $\{C_i^{(d,\underline{r})}\}_i$ and H_r is (in general not regularly) covered by $\{C_i^{(d,\bar{r})}\}_i$. See Figure 5.2 for an illustration in the two-dimensional case. This implies

$$\bigcup_{i=1}^{N(d,\underline{r})} F_{\mathrm{red}}^{-1}(C_i^{(d,\underline{r})}) = F_{\mathrm{red}}^{-1}(\bigcup_{i=1}^{N(d,\underline{r})} (C_i^{(d,\underline{r})})) \subset F_{\mathrm{red}}^{-1}(H_r), \qquad (5.35)$$

$$F_{\mathrm{red}}^{-1}(H_r) \subset F_{\mathrm{red}}^{-1}(\bigcup_{i=1}^{N(d,\bar{r})} (C_i^{(d,\bar{r})})) = \bigcup_{i=1}^{N(d,\bar{r})} F_{\mathrm{red}}^{-1}(C_i^{(d,\bar{r})}). \qquad (5.36)$$

By definition of M_1, M_2,

$$\frac{d}{2M_1} \le \frac{d\sqrt{2}}{2} \le \frac{dM_2}{2}.$$

By (i) together with the assumption $\frac{dM_2}{2} \le \delta$, we get by the first part of the lemma for arbitrary $r > 0$

$$\lambda_\ell\big(F_{\mathrm{red}}^{-1}(C_i^{(d,r)})\big) = \lambda_\ell\big(F_{\mathrm{red}}^{-1}(B(c_i, \tfrac{d}{2}))\big) = \lambda_\ell\big(F_{\mathrm{red}}^{-1}(\{c_i\}^{+\frac{d}{2}})\big)$$
$$\le \lambda_\ell\big(\{F_{\mathrm{red}}^{-1}(c_i)\}^{+\frac{dM_2}{2}}\big) = (dM_2)^\ell$$

and accordingly

$$\left(\frac{d}{M_1}\right)^\ell = \lambda_\ell\big(\{F_{\mathrm{red}}^{-1}(c_i)\}^{+\frac{d}{2M_1}}\big) \le \lambda_\ell\big(F_{\mathrm{red}}^{-1}(\{c_i\}^{+\frac{d}{2}})\big)$$
$$= \lambda_\ell\big(F_{\mathrm{red}}^{-1}(B(c_i, \tfrac{d}{2}))\big) = \lambda_\ell\big(F_{\mathrm{red}}^{-1}(C_i^{(d,r)})\big).$$

We use the monotonicity of the function $r \mapsto N(d,r)$ and the estimate

$$2r - d \le \underline{r} \le \bar{r} \le 2r + d$$

to get rid of the Gaussian brackets. The bounds of the second part of the lemma follow by (5.35) and (5.36):

$$M_1^{-\ell}\left((2r+d)^\ell - (2r-d)^\ell\right) = N(d, 2r-d)\left(\frac{d}{M_1}\right)^\ell \le \lambda_\ell\big(F_{\text{red}}^{-1}(H_r)\big),$$

$$\lambda_\ell\big(F_{\text{red}}^{-1}(H_r)\big) \le N(d, 2r+d)(dM_2)^\ell = M_2^\ell\left((2r+3d)^\ell - (2r+d)^\ell\right).$$

$\square$

Remark. *Observe that there is an obvious relation between Lemma 5.12* (ii) *and the theory of isoperimetric inequalities. If we write B_{r+d} as*

$$B_{r+d} = B_r + [-d,d]^\ell = \{x+y : x \in B_r, y \in [-d,d]^\ell\},$$

we get a lower bound of $\lambda_\ell(B_{r+d} \setminus B_r) = \lambda_\ell(B_{r+d}) - \lambda_\ell(B_r)$ by the so-called Brunn-Minkowsky inequality (see [Gar02]). In the notion of our context, it states

$$\lambda_\ell(B_{r+d})^{1/\ell} = \lambda_\ell(B_r + [-d,d]^\ell)^{1/\ell} \ge \lambda_\ell(B_r)^{1/\ell} + \lambda_\ell([-d,d]^\ell)^{1/\ell} = \lambda_\ell(B_r)^{1/\ell} + 2d. \tag{5.37}$$

This lower bound implies a lower bound in equation (5.34) *by the regularity of F_{red}. But while the Brunn-Minkowsky inequality holds for all convex bodies in $\mathbb{R}^\ell$, it is obvious that a reverse inequality, i.e. an upper bound in* (5.37), *is only valid in some special cases. See [BBP95] for details on reverse Brunn-Minkowski inequalities. Indeed, the regularity of the box collection $\mathcal{B}_k^{(\ell)}$ is essential in the lemma. For boxes with arbitrary radius vector, one cannot expect a proper upper bound in* (5.34) *relative to $\lambda_\ell(B_r)$.*

The main ingredient for the proof of Theorem 5.11 is the following embedding. It relates the N-dimensional Lebesgue measures in the entries of $P_{\text{I,I}}$ with the ℓ-dimensional Lebesgue measures in the entries of $P_k^{(\text{red})}$.

Lemma 5.13. *Let the assumptions of Theorem 5.11 hold. For $M_1, \delta > 0$ given as above let $0 < \varepsilon \le \delta$. Then*

$$\frac{1}{(2r)^\ell}\lambda_\ell\left(F_{\text{red}}^{-1}((B_i^{(\ell)})^{-M_1\varepsilon}) \cap B_j^{(\ell)}\right) \le p_{ij} \le \frac{1}{(2r)^\ell}\lambda_\ell\left(F_{\text{red}}^{-1}((B_i^{(\ell)})^{+M_1\varepsilon}) \cap B_j^{(\ell)}\right) \tag{5.38}$$

for every $1 \le i, j \le K$.

Proof. The result is based on the special shape of the boxes assumed in (5.27). For $i, j \in \{1, \dots, K\}$ let $u \in F^{-1}(B_i^{(N)}) \cap B_j^{(N)}$. Due to (5.27) we get

$$u = \begin{pmatrix} u_1 \\ u_2 \end{pmatrix}, \quad u_1 \in B_j^{(\ell)}, \ u_2 \in [-\varepsilon, \varepsilon]^{N-\ell}.$$

In particular using $\varepsilon \le \delta$, it follows that $u \in \mathcal{W}^{+\delta}$ and by the mean value theorem

$$\left\|F(u) - \begin{pmatrix} F_{\text{red}}(u_1) \\ 0 \end{pmatrix}\right\|_\infty \overset{(5.28)}{=} \left\|F(u) - F\begin{pmatrix} u_1 \\ 0 \end{pmatrix}\right\|_\infty \le M_1 \left\|u - \begin{pmatrix} u_1 \\ 0 \end{pmatrix}\right\|_\infty \le M_1\varepsilon. \tag{5.39}$$

With $F(u) \in B_i^{(N)} = B_i^{(\ell)} \times [-\varepsilon, \varepsilon]^{N_\ell}$ it follows

$$F_{\text{red}}(u_1) \in (B_i^{(\ell)})^{+M_1\varepsilon}. \tag{5.40}$$

By definition of p_{ij}, $i, j \in \{1, \ldots, K\}$, this implies:

$$(2r)^{\ell}(2\varepsilon)^{N-\ell} p_{ij} = \lambda_N(B_j^{(N)}) p_{ij} = \lambda_N\big(F^{-1}(B_i^{(N)}) \cap B_j^{(N)}\big) = \int_{B_j^{(N)}} \mathbb{1}_{B_i^{(N)}}(F(u))\, du$$

$$\overset{(5.40)}{\leq} \int_{B_j^{(N)}} \mathbb{1}_{\left(B_i^{(\ell)}\right)^{+M_1\varepsilon}}(F_{\mathrm{red}}(u_1))\, d\begin{pmatrix} u_1 \\ u_2 \end{pmatrix}$$

$$= \int_{B_j^{(\ell)}} \int_{[-\varepsilon,\varepsilon]^{N-\ell}} du_2\, \mathbb{1}_{\left(B_i^{(\ell)}\right)^{+M_1\varepsilon}}(F_{\mathrm{red}}(u_1))\, du_1$$

$$= (2\varepsilon)^{N-\ell} \lambda_\ell\big(F_{\mathrm{red}}^{-1}(B_i^{(\ell)})^{+M_1\varepsilon} \cap B_j^{(\ell)}\big).$$

Dividing by $(2r)^{\ell}(2\varepsilon)^{N-\ell}$ provides the second inequality in (5.38).

For the first inequality, consider $u \in B_j^{(N)} \setminus F^{-1}(B_i^{(N)})$. As above, $u \in B_j^{(N)}$ implies (5.39). By assumption $\mathbf{C_1}$, the box collection $\mathcal{B}_I$ is positive invariant, hence $\begin{pmatrix} v_1 \\ v_2 \end{pmatrix} :=$ $F(u) \notin B_i^{(N)}$ implies $v_1 \notin B_i^{(\ell)}$ and with (5.39) it follows

$$F_{\mathrm{red}}(u_1) \notin \big(B_i^{(\ell)}\big)^{-M_1\varepsilon}. \tag{5.41}$$

As above, we get

$$(2r)^{\ell}(2\varepsilon)^{N-\ell} p_{ij} = \lambda_N(B_j^{(N)}) p_{ij} = \lambda_N\big(F^{-1}(B_i^{(N)}) \cap B_j^{(N)}\big)$$

$$\overset{(5.41)}{\geq} \int_{B_j^{(N)}} \mathbb{1}_{\left(B_i^{(\ell)}\right)^{-M_1\varepsilon}}(F_{\mathrm{red}}(u_1))\, d\begin{pmatrix} u_1 \\ u_2 \end{pmatrix}$$

$$= (2\varepsilon)^{N-\ell} \lambda_\ell\big(F_{\mathrm{red}}^{-1}(B_i^{(\ell)})^{-M_1\varepsilon} \cap B_j^{(\ell)}\big).$$

Dividing by $\lambda_N(B_j^{(N)}) = (2r)^{\ell}(2\varepsilon)^{N-\ell}$ gives the first inequality in (5.38). Thus the lemma is proved. $\qquad\square$

Proof of Theorem 5.11. Let $P_{1,\mathrm{I}} = (p_{ij})_{ij}$, $P_k^{(\mathrm{red})} = (p_{ij}^{(\mathrm{red})})_{ij} \in \mathbb{R}^{K,K}$ be given with the assumptions as above and $0 < \varepsilon \leq \varepsilon_0$. In particular,

$$\varepsilon \leq \varepsilon_0 \leq \frac{2\delta}{M_1 M_2} \leq \frac{2\delta}{2} = \delta. \tag{5.42}$$

We derive an error bound for $p_{ij} - p_{ij}^{(\mathrm{red})}$ where p_{ij} is defined by the N-dimensional and $p_{ij}^{(\mathrm{red})}$ by the ℓ-dimensional Lebesgue measure. By (5.42) we can write p_{ij} in terms of the ℓ-dimensional Lebesgue measure according to Lemma 5.13. Hence we get an approximation for the difference $p_{ij} - p_{ij}^{(\mathrm{red})}$ in terms of ℓ-dimensional Lebesgue measures. Using this representation, we can derive an estimate for the difference in terms of the Lebesgue measure of an ℓ-dimensional box hull such that Lemma 5.12 ii) applies and we get

$$(2r)^{\ell}\big(p_{ij} - p_{ij}^{(\mathrm{red})}\big) = \lambda_\ell(B_j^{(\ell)})\big(p_{ij} - p_{ij}^{(\mathrm{red})}\big) \tag{5.43}$$

$$\in \Big[\lambda_\ell\Big(F_{\mathrm{red}}^{-1}((B_i^{(\ell)})^{-M_1\varepsilon}) \cap B_j^{(\ell)}\Big) - \lambda_\ell\Big(F_{\mathrm{red}}^{-1}(B_i^{(\ell)}) \cap B_j^{(\ell)}\Big),$$

$$\lambda_\ell\Big(F_{\mathrm{red}}^{-1}((B_i^{(\ell)})^{+M_1\varepsilon}) \cap B_j^{(\ell)}\Big) - \lambda_\ell\Big(F_{\mathrm{red}}^{-1}(B_i^{(\ell)}) \cap B_j^{(\ell)}\Big)\Big]$$

$$
\begin{aligned}
&= \left[-\lambda_\ell \left(F_{\text{red}}^{-1} \left(B_i^{(\ell)} \setminus (B_i^{(\ell)})^{-M_1\varepsilon} \right) \cap B_j^{(\ell)} \right), \right. \\
&\qquad \left. \lambda_\ell \left(F_{\text{red}}^{-1} \left((B_i^{(\ell)})^{+M_1\varepsilon} \setminus B_i^{(\ell)} \right) \cap B_j^{(\ell)} \right) \right] \\
&\subset \left[-\lambda_\ell \left(F_{\text{red}}^{-1} \left(B_i^{(\ell)} \setminus (B_i^{(\ell)})^{-M_1\varepsilon} \right) \right), \lambda_\ell \left(F_{\text{red}}^{-1} \left((B_i^{(\ell)})^{+M_1\varepsilon} \setminus B_i^{(\ell)} \right) \right) \right] \\
&\overset{(5.34)}{\subset} \left[-M_1^\ell \left((2r + M_1\varepsilon)^\ell - (2r - M_1\varepsilon)^\ell \right), \right. \\
&\qquad \left. M_2^\ell \left((2r + 3M_1\varepsilon)^\ell - (2r + M_1\varepsilon)^\ell \right) \right].
\end{aligned}
\tag{5.44}
$$

By convexity of $a \mapsto a^\ell$ the upper bound dominates the negative of the lower bound and we get

$$
|p_{ij} - p_{ij}^{(\text{red})}| \leq \left(\frac{M_2}{2r} \right)^\ell \left((2r + 3M_1\varepsilon)^\ell - (2r + M_1\varepsilon)^\ell \right)
$$

By the mean value theorem on $a \mapsto a^\ell$ it follows with $\varepsilon \leq \varepsilon_0 \leq \alpha \frac{2r}{3M_1}$

$$
\begin{aligned}
|p_{ij} - p_{ij}^{(\text{red})}| &\leq 2\ell M_1 \varepsilon \left(\frac{M_2}{2r} \right)^\ell (2r + 3M_1\varepsilon)^{\ell-1} \\
&\leq \varepsilon\, 2\ell M_1 \left(\frac{M_2}{2r} \right)^\ell (2r)^{\ell-1} (1 + \alpha)^{\ell-1} \\
&= \varepsilon \frac{\ell M_1}{r} (1 + \alpha)^{\ell-1} M_2^\ell.
\end{aligned}
\tag{5.45}
$$

The norm-wise estimate (5.32) for $P_{1,\mathrm{I}} - P_k^{(\text{red})}$ follows from (5.31) by definition of the induced matrix norm $\| \cdot \|_\infty$:

$$
\| P_{1,\mathrm{I}} - P_k^{(\text{red})} \|_\infty = \max_{1 \leq i \leq K} \sum_{j=1}^{K} |p_{ij} - p_{ij}^{(\text{red})}| \overset{(5.31)}{\leq} \varepsilon \frac{K \ell M_1}{r} (1 + \alpha)^{\ell-1} M_2^\ell.
$$

$\square$

Remark. *In step (5.44) we use a quite coarse estimate as we ignore the intersection of the hulls with the boxes $B_j^{(\ell)}$. Thus it seems worthwhile to look for a more suitable matrix norm than the one induced by the maximum norm. A good choice is the norm $\| \cdot \|_S$ on $\mathbb{R}^{K,K}$ defined by*

$$
\| A \|_S = \max_{i=1,\dots,K} \left(\max_{n=1,\dots,K} | \sum_{j=1}^{n} a_{ij} | \right).
$$

An easy calculation shows that $\| \cdot \|_S$ can be written as an induced matrix norm $\| \cdot \|_{1,-1}$ with the sum norm $\| \cdot \|_1$ and the Spijker norm $\| v \|_{-1} := \max_{n=1,\dots,K} | \sum_{j=1}^{n} v_j |$ via

$$
\| A \|_S = \| A^T \|_{1,-1} = \sup_{\| v \|_1 \leq 1} \| A^T v \|_{-1}.
$$

In this norm, the estimate for the distance between $P_{1,\mathrm{I}}$ and $P_k^{(\text{red})}$ is

$$
\| P_{1,\mathrm{I}} - P_k^{(\text{red})} \|_S \leq \varepsilon \frac{\ell M_1}{r} (1 + \alpha)^{\ell-1} M_2^\ell.
$$

This follows from an approximation of

$$\sum_{j=1}^{n}(2r)^{\ell}(p_{ij} - p_{ij}^{(\text{red})})$$

instead of $(2r)^{\ell}(p_{ij} - p_{ij}^{(\text{red})})$ *in* (5.43).

However, it is not clear if the bound in $\|\cdot\|_S$ implies a proper bound for the corresponding Perron roots. For the perturbation result (5.16) we need a bound in the maximum norm as given by (5.32) that implies the following concluding corollary.

Corollary 5.14. *Let a discrete dynamical system be given by a diffeomorphism* $F : \mathbb{R}^N \to \mathbb{R}^N$ *and let the reduced system be defined by* $F_{\text{red}} : \mathbb{R}^\ell \to \mathbb{R}^\ell$. *Further, assume that the POD subspace and the box collections satisfy assumptions* $\mathbf{C_1}$, $\mathbf{C_2}$. *If* $P_k^{(\text{red})}$ *is irreducible, the discrete invariant measures* μ_k, $\mu_k^{(\text{red})}$ *satisfy the following properties:*

- μ_k *is defined by a vector* $u = \begin{pmatrix} u_{\text{I}} \\ 0 \end{pmatrix} \in \mathbb{R}^{K_N}$ *with* $u_{\text{I}} \in \mathbb{R}^K$, *while* $\mu_k^{(\text{red})}$ *is defined by a vector* $u_k^{(\text{red})} \in \mathbb{R}^K$. *In particular* $\text{supp}(\mu_k) \in \mathcal{B}_{\text{I}}$.

- *The distance of the discrete measures is bounded by*

$$\|u_{\text{I}} - u_k^{(\text{red})}\|_\infty \leq \varepsilon \frac{K\ell M_1}{r}(1 + \alpha)^{\ell-1} M_2^\ell \bar{\kappa}(P_k^{(\text{red})})$$
$$\leq \varepsilon \frac{K\ell M_1}{r}(1 + \alpha)^{\ell-1} M_2^\ell \frac{2(\ell - 1)}{\chi(P_k^{(\text{red})})} \tag{5.46}$$

where $\bar{\kappa}(P)$ *is the condition number and* $\chi(P) \in (0, K]$ *the character of a Perron-Frobenius matrix* P *defined by* (5.17) *and* (5.19).

5.3.3   Combination with POD error

By assumptions $\mathbf{C_1}$ and $\mathbf{C_2}$, the POD subset $\mathcal{W}$ is F-invariant and attracting. This implies that the whole dynamic of the system is located in $\mathcal{W}$. In more realistic cases, the global attractor is not a subset of a low-dimensional linear subspace. Thus we cannot expect that $\mathcal{W}$ satisfies the strong assumptions $\mathbf{C_1}$, $\mathbf{C_2}$. Nevertheless it is reasonable to assume that the POD model reduction defines a subspace approximating the relevant dynamical sets of F within a tolerance. Thus we weaken the assumptions of F-invariance and attraction of $\mathcal{W}$ by the following:

$$\|F(u_1, 0) - (F_{\text{red}}(u_1), 0)\| \leq C_{\mathcal{W}} \quad \text{for all } u_1 \in \mathcal{W}. \tag{5.47}$$

where we assume $C_{\mathcal{W}} > 0$ to be small. $\mathcal{B}_{\text{I}} = \mathcal{B}_{\text{I}}(\varepsilon)$ is now assumed to be positive invariant and attracting.

Recall that typical error estimates of the POD method treat the case of single trajectories, see the result by Kunisch and Volkwein in Theorem 2.9. Roughly speaking, for single trajectories $u(t)$ the error bound is given by

$$C_{\mathcal{W}} = C \sum_{i=\ell+1}^{\infty} \sigma_i^2$$

where σ_i are the singular values of the matrix of snapshots $y_j = u(jT)$ with a given stepsize $T > 0$.

It is still reasonable to assume (5.47) if the POD method is based on snapshots built from many short time trajectories instead of a single long trajectory. See the discussion at the end of Chapter 2 for details.

Let us now assume (5.47) instead of invariance of $\mathcal{W}$. We prove a slight modification of Lemma 5.12. For $u \in B_j^{(N)}$ we get the estimate

$$\left\| F(u) - \begin{pmatrix} F_{\text{red}}(u_1) \\ 0 \end{pmatrix} \right\|_\infty \leq \left\| F(u) - F\begin{pmatrix} u_1 \\ 0 \end{pmatrix} \right\|_\infty + \left\| F\begin{pmatrix} u_1 \\ 0 \end{pmatrix} - \begin{pmatrix} F_{\text{red}}(u_1) \\ 0 \end{pmatrix} \right\|_\infty$$
$$\leq M_1 \varepsilon + C_{\mathcal{W}} \tag{5.48}$$

instead of (5.39). For $u \in B_j^{(N)} \cap F^{-1}(B_i^{(N)})$, we get

$$F_{\text{red}}(u_1) \in \left(B_i^{(\ell)} \right)^{+(M_1\varepsilon + C_{\mathcal{W}})}$$

as in the proof of Lemma 5.12. For $u \in B_j^{(N)} \setminus F^{-1}(B_i^{(N)})$ we get by positive invariance of $\mathcal{B}_I$:

$$F_{\text{red}}(u_1) \notin \left(B_i^{(\ell)} \right)^{-(M_1\varepsilon + C_{\mathcal{W}})}.$$

In summary we get the following result similar to (5.38)

$$\frac{1}{(2r)^\ell} \lambda_\ell \left(F_{\text{red}}^{-1}((B_i^{(\ell)})^{-(M_1\varepsilon + C_{\mathcal{W}})}) \cap B_j^{(\ell)} \right) \leq p_{ij},$$
$$p_{ij} \leq \frac{1}{(2r)^\ell} \lambda_\ell \left(F_{\text{red}}^{-1}((B_i^{(\ell)})^{+(M_1\varepsilon + C_{\mathcal{W}})}) \cap B_j^{(\ell)} \right). \tag{5.49}$$

The remaining parts of the proof of Theorem 5.11 carry over. For given $\alpha > \frac{3C_{\mathcal{W}}}{2r}$ and for every $\varepsilon > 0$ with

$$\varepsilon < \tilde{\varepsilon}_0 = \min \left(\frac{\alpha 2r - 3C_{\mathcal{W}}}{3M_1}, \frac{2\delta}{M_1 M_2} \right)$$

we get an entry-wise bound similar to (5.31):

$$|p_{ij} - p_{ij}^{(\text{red})}| \leq (\varepsilon M_1 + C_{\mathcal{W}}) \frac{\ell}{r} (1 + \alpha)^{\ell-1} M_2^\ell. \tag{5.50}$$

An estimate corresponding to (5.32) for the maximum norm and an estimate for the Perron roots corresponding to (5.46) follow accordingly.

Chapter 6

Numerical Experiments

In this chapter, we analyze the reduced-space algorithms derived in Chapter 3 in various numerical experiments. We compute discrete measures in two dynamical systems. One is given by an embedding of the Lorenz system into a high-dimensional state space. Our main application is the second system given by a fully discretized Chafee-Infante problem.

Our numerical experiments partly support the main results of the preceding chapters and partly indicate open problems. For the embedded Lorenz system, we have a closer look at the POD computation step of the reduced-space algorithms. For different numbers and lengths of snapshots, we investigate the approximation of the relevant transformation space by the computed POD modes. The results illustrate the considerations of Section 2.3. For the Chafee-Infante problem, we analyze the distances between the discrete measures of the AIM and the PODAIM algorithms, and the effect of varying the POD dimension. These experiments are strongly related to the error results derived in Chapter 5.

We end the chapter with a test of the PODADAPT algorithm. It turns out that the adaptation of the POD space works well in the test example if the parameters of the algorithm are set up properly. But the additional computational effort does not yield better approximation results than the PODAIM algorithm with fixed POD space.

6.1 Representing discrete measures in high dimensions

First, we develop a suitable representation of discrete measures in high dimensions in order to present the results of the algorithms described in Chapter 3. The dimension $N \in \mathbb{N}$ of the state space is in general too high for a direct plot of the densities of discrete measures $\mu_k : \mathcal{B}(\mathbb{R}^N) \to [0, 1]$.

The main applications of the algorithms are discretizations of scalar parabolic problems with linear finite elements. Thus a state $u \in \mathbb{R}^N$ corresponds to a polygonal curve with vertices

$$h_u := \left\{ (0, 0), \left(\frac{1}{N+1}, u_1 \right), \dots, \left(\frac{N}{N+1}, u_N \right), (1, 0) \right\}.$$

It is an obvious approach to use this identification for the discrete measures as well. Therefore, we represent a discrete measure $\mu = \mu_k^{(\mathrm{pod})} \in \mathcal{M}^1(\mathbb{R}^N)$ derived by the PODAIM algorithm 3.1 by a density function

$$h_\mu : \mathcal{Q} \to [0, 1].$$

Here, $\mathcal{Q} \subset \mathcal{B}([0,1] \times \mathbb{R})$ denotes a collection of boxes according to the discretization.

We now describe the construction of h_μ corresponding to a discrete measure $\mu = \mu_k^{(\text{pod})}$ on a box collection $\mathcal{B}_k \subset B_0^{(N)}$ with starting box $B_0^{(N)} = B(c_0, r_0)$. With $r := \|r_0\|_\infty$ define the box collection $\mathcal{Q}$ by

$$\mathcal{Q} := \{Q_{ij} = R_i \times S_j : 1 \le i \le N, 1 \le j \le J\},$$

$$R_i = \left[\frac{2i-1}{2N+2}, \frac{2i+1}{2N+2} \right], \quad S_j = \left[r\left(-1 + \frac{2j-2}{J} \right), r\left(-1 + \frac{2j}{J} \right) \right].$$

Recall the geometric definition (3.6) of a PODAIM measure on the box collection $\mathcal{B}_k = \{B_1, \ldots, B_K\}$ with boxes $B_\alpha = B(c_\alpha, r_\alpha) \subset \mathbb{R}^\ell$:

$$\mu(A) = \mu_k^{(\text{pod})}(A) = \sum_{\alpha=1}^K u_\alpha \frac{\lambda_\ell(B_\alpha \cap W_1^{-1}(A))}{\lambda_\ell(B_\alpha)}.$$

Following the idea to view states as polygonal curves on $[0,1]$, we count those support boxes of our discrete measure whose i-th center component is located in an interval S_j defined above. This leads to

$$h_\mu(Q_{ij}) := \sum_{\alpha=1}^K u_\alpha \mathbb{1}_{Q_{ij}}\left(\frac{i}{N+1}, (W_1 c_\alpha)_i \right) = \sum_{\alpha=1}^K u_\alpha \mathbb{1}_{S_j}\left((W_1 c_\alpha)_i \right).$$

By this definition, h_μ approximates the so-called *marginal distributions* of μ. In probability theory, a marginal distribution is given by the distribution in a fixed coordinate $1 \le i \le N$ where the other coordinates are integrated (see [Bil99], 1.2). Here, $h_\mu(Q_{ij})$ approximates the measure of the strip

$$A_{ij} := \mathbb{R}^{i-1} \times S_j \times \mathbb{R}^{N-i}$$

in the following sense:

$$\mu(A_{ij}) = \sum_{\alpha=1}^K u_\alpha \frac{\lambda_\ell\left(B_\alpha \cap W_1^{-1}(A_{ij})\right)}{\lambda_\ell(B_\alpha)}$$

$$= \sum_{\alpha=1}^K u_\alpha \frac{\lambda_\ell(\{x \in B_\alpha : (W_1 x)_i \in S_j\})}{\lambda_\ell(B_\alpha)}$$

$$\approx \sum_{\alpha=1}^K u_\alpha \begin{cases} 1, & \text{if } (W_1 c_\alpha)_i \in S_j \\ 0, & \text{otherwise} \end{cases}$$

$$= \sum_{\alpha=1}^K u_\alpha \mathbb{1}_{S_j}\left((W_1 c_\alpha)_i \right) \; = \; h(Q_{ij}).$$

The approximation process is realistic since we can assume that after some recursion steps

$$\text{diam}(B_\alpha) = 2\|r_\alpha\|_\infty \ll 2\frac{\|r_0\|_\infty}{J} = \text{diam}(S_j).$$

By this property most of the projected boxes $W_1(B_\alpha)$ either lie completely in A_{ij} or completely in its complement and the test with the centers is reasonable.

With this discrete marginal operator $\mu \mapsto h_\mu$ we visualize our results by color-coding Q_{ij} according to the value $h_\mu(Q_{ij})$. It is easy to see that the corresponding matrix $H_\mu = (h_\mu(Q_{ij}))_{ij} \in \mathbb{R}^{N,J}$ is (row-)stochastic if the support boxes are located in $B_0^{(N)}$:

$$\sum_{j=1}^{J} h(Q_{ij}) = \sum_{\alpha=1}^{K} u_\alpha \sum_{j=1}^{J} \mathbb{1}_{Q_{ij}} \left(\frac{i}{N+1}, (W_1 c_\alpha)_i \right)$$
$$= \sum_{\alpha=1}^{K} u_\alpha = 1.$$

6.2 Embedded Lorenz system

We start our numerical experiments with a dynamical system arising from an ordinary differential equation based on the well-known Lorenz system ([Lor63]). We use this three-dimensional equation to analyze the PODAIM algorithm 3.1 and in particular its model reduction step.

Therefore in the following, we define a discrete dynamical system $v_{i+1} = G_h(v_i)$ as a slightly perturbed embedding of the 3-dimensional Lorenz system into a phase space of dimension $N > 3$. We start by adding coordinates with trivial dynamics to the Lorenz system and get an N-dimensional system of differential equations

$$u' = F_L(u), \quad \text{with } F_L : \mathbb{R}^N \to \mathbb{R}^N, \tag{6.1}$$

$$F_L(u) = \begin{pmatrix} \sigma(u_2 - u_1) \\ \rho u_1 - u_2 - u_1 u_3 \\ u_1 u_2 - \beta u_3 \\ -\alpha u_4 \\ \vdots \\ -\alpha u_N \end{pmatrix}. \tag{6.2}$$

Observe that the first three coordinates in (6.2) simply define the Lorenz equation. The other coordinates of a solution to (6.1) decay exponentially if we assume $\alpha > 0$.

The differential equation (6.1) defines the following discrete dynamical system by use of Euler's method:

$$u^{(i+1)} = F_h(u^{(i)}), \quad F_h(v) = v + hF_L(v) \tag{6.3}$$

We define a quadratic perturbation of (6.3) via the scalar function $t_\varepsilon : \mathbb{R} \to \mathbb{R}$ given by

$$t_\varepsilon(x) = (1 - \varepsilon)x + \varepsilon x^2.$$

Some analysis shows that t_ε is one-to-one from $D_\varepsilon := \left(-\frac{1-\varepsilon}{2\varepsilon}, \infty \right)$ onto $R_\varepsilon := \left(-\frac{(1-\varepsilon)^2}{4\varepsilon^2}, \infty \right)$. Thus the inverse function $t_\varepsilon^{-1} : R_\varepsilon \to D_\varepsilon$ exists with

$$t_\varepsilon^{-1}(y) = \sqrt{\frac{y}{\varepsilon} + \frac{(1-\varepsilon)^2}{4\varepsilon^2}} - \frac{1-\varepsilon}{2\varepsilon}.$$

By use of the scalar quadratic function t_ε we define a transformation of diagonal shape by

$$(T_\varepsilon(v))_i = t_\varepsilon(v_i), \quad i = 1, \ldots, N.$$

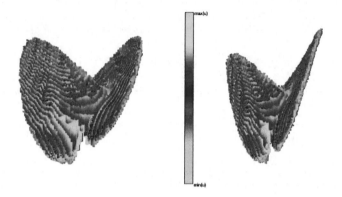

Figure 6.1: Resulting box collection of PODAIM algorithm for the embedded Lorenz system in $N = 10$ dimensions from two different viewing angles.

Obviously T_ε is a bijection with $(T_\varepsilon^{-1}(v))_i = t_\varepsilon^{-1}(v_i)$. Finally, by use of a randomly chosen orthogonal matrix $Q = \mathrm{col}(q_1, \ldots, q_N) \in \mathbb{R}^{N,N}$, the orientation of the quadratic perturbation defined by T_ε is randomized. With these preparations, we set

$$v_{i+1} = G_{h,\varepsilon}(v_i) := T_\varepsilon(QF_h(Q^T T_\varepsilon^{-1}(v_i))), \quad i = 1, \ldots, N. \tag{6.4}$$

6.2.1   Solution of the PODAIM algorithm

In Figure 6.1 we present the result of the PODAIM algorithm for the system (6.4) in dimension $N = 10$ for the following parameter values:

$$\sigma = 10, \qquad \rho = 28, \qquad \beta = 8/3,$$
$$\varepsilon = 0.001, \qquad \alpha = 0.9, \qquad h = 0.01.$$

For the POD step 3.1 we compute $m = 100$ trajectories of length $T = 1\,000$ starting with randomly chosen initial data in the unit cube. This results in the following singular values

$$\sigma_1 = 742.0957, \qquad \sigma_2 = 333.8305, \qquad \sigma_3 = 75.0522, \qquad \sigma_4 = 2.3911.$$

Due to the gap in magnitude between σ_3 and σ_4 the POD dimension $\ell = 3$ was chosen.

By definition of the discrete dynamical system, we expect that the POD model reduction detects a subspace located near the subspace spanned by the first three columns $q_1, q_2, q_3 \in \mathbb{R}^{10}$ of the linear transformation Q, since the quadratic perturbation added to the Lorenz system is of small scale.

We test this hypothesis by evaluating the canonical angle $\angle(W_1, Q_1)$ where W_1 represents the POD space computed by the PODAIM algorithm and $Q_1 = \mathrm{col}(q_1, q_2, q_3)$. See

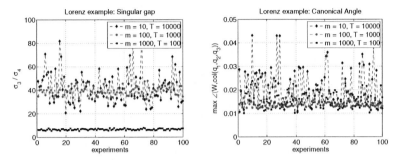

Figure 6.2: Computation of the POD space for the embedded Lorenz system: Singular gap $\frac{\sigma_3}{\sigma_4}$ and canonical angle with transformation space for different choices of the parameters $m, T \in \mathbb{N}$.

parameters (m, T)	$(10, 10\,000)$	$(100, 1\,000)$	$(1\,000, 100)$	$(10\,000, 10)$
average singular gap $\frac{\sigma_3}{\sigma_4}$	42.0180	39.1220	6.5499	1.0059
average angle $\angle(W_1, Q_1)$	0.0206	0.0137	0.0160	1.5702

Table 6.1: Average values for the experiments plotted in Figure 6.2.

Definition A.11 in the appendix for the definition of the canonical angle already used in the analysis of the POD method in Chapter 2. In our numerical experiment we get

$$\angle(W_1, Q_1) = \mathrm{diag}(0.0009, 0.0034, 0.0111). \tag{6.5}$$

Hence, the POD model reduction works as expected in our model example.

Figure 6.1 shows the resulting box collection in the POD space after $k = 39$ recursion steps, i.e. 13 bisections in each space dimension. The support of the discrete measure consists of $K = 348\,954$ boxes. The support boxes are colored according to the discrete measure on a logarithmic scale as suggested in [DHJR97].

If we compare the discrete measure in the figure with the known results for the Lorenz system and the AIM algorithm in [DHJR97], we see that the support apparently approximates an object with a shape similar to the well-known Lorenz attractor. This indicates that no relevant dynamics are lost due to the model reduction process.

6.2.2 Analysis of the POD computation step

We use our test example for a discussion of how to choose the collection of snapshots in the POD computation step 3.1. Recall that the collection of snapshots is computed by short trajectories of length $T \in \mathbb{N}$ for $m \in \mathbb{N}$ randomly chosen initial points. In Figure 6.2, we analyze the POD computation for collections of snapshots based on 3 different parameter pairs (m, T). By keeping $mT = 10^5$ constant, the computational effort remains roughly the same throughout the experiments. Note that the POD computation depends on the distribution of the initial points. Therefore, we have repeated the experiments for each choice of parameters. See Table 6.1 for the average values of Figure 6.2.

In the first plot of Figure 6.2 we see that the spectral gap between the third and fourth singular value is distinctive for trajectories of a proper length, see the curves for $(m, T) =$

$(100, 1\,000)$ and $(m, T) = (10, 10\,000)$. For the parameter choice $(m, T) = (1\,000, 100)$, the spectral gap is very low. Hence, the critical limit for T in this example is located somewhere between $T = 10^2$ and $T = 10^3$. If the length of trajectories falls below this limit, the spectral gap is too small to provide the proper POD dimension.

The obvious next experiment is given by the parameters $(m, T) = (10\,000, 10)$. The corresponding collection of snapshots is not evaluated in the plot, since the ratio $\frac{\sigma_3}{\sigma_4}$ is practically given by 1 in this case, see Table 6.1. Hence the proper POD dimension $\ell = 3$ cannot be detected automatically by the algorithm.

For the same choices of snapshots as in the first plot we illustrate the maximal canonical angle

$$\max(\angle(W_1, Q_1))$$

during the experiments in the second plot of Figure 6.2. Obviously, in all three cases the subspace given by Q_1 is detected with similar precision in Step 3.1 of the algorithm. The example with only $m = 10$ trajectories results in a POD space with a slightly larger canonical angle. The example $(m, T) = (10\,000, 10)$ mentioned above is again not plotted in the figure. In this case the POD space $W_1 = W_1(m, T)$ does not detect the linear transformation, $\max(\angle(W_1, Q_1)) \approx \frac{\pi}{2}$.

These results shed some light on the problem of balancing m and T which we already discussed in Section 2.3.6. Compared to the analysis shown in Figure 2.6, this example is more realistic in the sense that the system possesses a nontrivial attractor. As proposed in Section 2.3.6, there is a trade-off between the number of trajectories and the length of trajectories when we look for an optimal collection of snapshots for the POD method.

6.3   The Chafee-Infante problem

As a main application of our algorithm, we look at a parabolic system given by the scalar Chafee-Infante problem with a cubic nonlinearity:

$$u_t = u_{xx} + \lambda(u - u^3), \quad x \in \Omega := (0, 1), \ t > 0, \tag{6.6}$$
$$u(0, t) = u(1, t) = 0, \qquad\qquad t > 0.$$

We sum up some theoretical results for this system before we discretize to a system of type (3.1).

6.3.1   Long-time behavior

The dynamical behavior of the Chafee-Infante problem is well-analyzed, e.g. in [Hen81], [Rob01], [Hal88] and [BV92]. We collect the most important results concerning the long-time behavior:

- There is an a-priori bound for solutions to initial values in the Sobolev space $H_0^1(\Omega)$. Hence (6.6) defines a continuous-time dynamical system $S : \mathbb{R}_+ \times H_0^1(\Omega) \to H_0^1(\Omega)$.

- The Chafee-Infante problem is a *gradient system*, i.e. it has a Lyapunov function $\mathcal{L} : H_0^1(\Omega) \to \mathbb{R}$. This function is explicitly given by

$$\mathcal{L}(u) = \int_\Omega \frac{1}{2}|\nabla u| - \frac{1}{2}u^2 + \frac{1}{4}u^4 \, du.$$

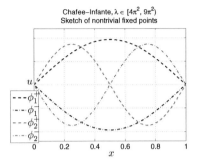

Figure 6.3: Chafee-Infante problem. Sketch of the four nontrivial fixed points $\{\phi_1^\pm, \phi_2^\pm\}$ for a parameter value $\lambda \in [4\pi^2, 9\pi^2)$.

- The dynamical system defined by (6.6) is *dissipative*, i.e. there is an absorbing set in $H_0^1(\Omega)$.

- The bifurcation points of (6.6) are given by the monotone sequence $(\lambda_n)_n$ with $\lambda_n = n^2\pi^2$. The set of fixed points $\mathbb{G}_\lambda$ depending on the parameter is discrete and given by the following theorem:

Theorem 6.1. *For given parameter $\lambda > 0$ choose $n \in \mathbb{N}$ such that $\lambda \in [\lambda_n, \lambda_{n+1})$. Then the Chafee-Infante problem possesses $2n + 1$ fixed points in $H_0^1(\Omega)$:*

$$\mathbb{G}_\lambda = \{\phi_0, \phi_1^\pm(\lambda), \dots, \phi_n^\pm(\lambda)\} \subset H_0^1(\Omega)$$

with the following properties

- $\phi_0 = 0$, $\phi_k^+(\lambda) = -\phi_k^-(\lambda)$ *with* $\frac{d}{dx}[\phi_k^+(\lambda)](0) > 0$ *for every* $1 \leq k \leq n$.

- *For every* $1 \leq k \leq n$ *the fixed point* $\phi_k^\pm(\lambda)$ *has* $k - 1$ *roots in* $\Omega = (0,1)$ *given by*

$$\left\{\frac{1}{k}, \frac{2}{k}, \dots, \frac{k-1}{k}\right\}.$$

- *For $n = 0$ the only fixed point $\phi_0 = 0$ is stable. For $n \geq 1$ the only stable fixed points are given by $\phi_1^\pm(\lambda)$.*

See Figure 6.3 for a sketch of the four nontrivial fixed points $\{\phi_1^\pm, \phi_2^\pm\}$ of the Chafee-Infante system for parameter values given by $4\pi^2 \leq \lambda < 9\pi^2$. We will focus on this parameter interval in the following numerical experiments.

The dissipation of the dynamical system defined by (6.6) implies the existence of a global attractor $\mathcal{A} \subset H_0^1(\Omega)$. The following theorem gives a detailed description of it:

Theorem 6.2 ([Rob01], Theorem 10.13). *Suppose that $S : \mathbb{R}_+ \times X \to X$ has a Lyapunov function on the attractor $\mathcal{A} \subset X$ and denote by $\mathbb{G}$ the set of fixed points of S. Then the attractor is given by the unstable manifold of $\mathbb{G}$: $\mathcal{A} = W^u(\mathbb{G})$.*

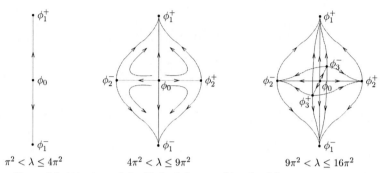

$$\pi^2 < \lambda \le 4\pi^2 \qquad\qquad 4\pi^2 < \lambda \le 9\pi^2 \qquad\qquad 9\pi^2 < \lambda \le 16\pi^2$$

Figure 6.4: Attractors of the Chafee-Infante problem for different parameter values

Furthermore, if X is connected and $\mathbb{G}$ is discrete, then the attractor can be written as

$$\mathcal{A} = W^u(\mathbb{G}) = \bigcup_{v \in \mathbb{G}} W^u(v)$$

and also as $\quad \mathcal{A} = W^s(\mathbb{G}) = \bigcup_{v \in \mathbb{G}} W^s(v).$

Figure 6.4 shows a sketch of the global attractor's shape for the Chafee-Infante problem with different parameter values $\lambda \ge 0$ inspired by the pictures in [Hen81].[1]

6.3.2 Discretization of the Chafee-Infante problem

In order to discretize in space, we choose a standard finite elements ansatz with linear basis functions as used already in Section 2.3.4 for a linear parabolic equation. Again, let $x_i = ih$, $i = 1, \ldots, N$ be equally distributed grid points in the unit interval with step size $h = \frac{1}{N+1}$. The set of linear finite elements is spanned by the hat functions $\Lambda_j = \Lambda_j(h) : [0,1] \to \mathbb{R}$, $j = 1, \ldots, N$, described by

$$\Lambda_j(x_i) = \delta_{ij} \quad \text{for all } i, j \in \{1, \ldots, N\}.$$

The finite element solution in $V_h = \operatorname{span}\{\Lambda_1, \ldots, \Lambda_N\}$ is given by the following weak formulation:

$$(\frac{d}{dt} u_h(t), \Lambda_j)_2 + a(u_h(t), \Lambda_j) = \lambda(u_h(t) - u_h(t)^3, \Lambda_j)_2, \quad \text{for all } 1 \le j \le N.$$

The inner products $(\cdot, \cdot)_2$ and $a(\cdot, \cdot)$ are given as in the linear case by (2.59). Analogously, for the coefficients of

$$u_h(t) = \sum_{i=1}^{N} \mathbf{u}_i(t) \Lambda_i \in V_h,$$

we get a system of ordinary differential equations

$$M_h \mathbf{u}'(t) + S_h \mathbf{u}(t) = \lambda G_h(\mathbf{u}(t)). \tag{6.7}$$

---

[1]Sincere thanks to Denny Otten for providing the figure from his diploma thesis [Ott09].

The mass matrix $M_h = \frac{h}{6}M$ and the stiffness matrix $S_h = \frac{1}{h}S$ are defined in (2.61). The nonlinear function $G_h : \mathbb{R}^N \to \mathbb{R}^N$ is defined by

$$G_h(\mathbf{u})_j = \int_0^1 \Lambda_j(x) \left(\sum_{i=1}^N \mathbf{u}\Lambda_i(x) - (\sum_{i=1}^N \mathbf{u}\Lambda_i(x))^3 \right) dx$$

$$= hG(\mathbf{u})$$

where $G : \mathbb{R}^N \to \mathbb{R}^N$ is given by

$$G(\mathbf{u})_j = \frac{\mathbf{u}_{j-1}}{6} + \frac{2\mathbf{u}_j}{3} + \frac{\mathbf{u}_{j+1}}{6} - \frac{\mathbf{u}_{j-1}^3}{20} - \frac{2\mathbf{u}_j^3}{5} - \frac{\mathbf{u}_{j+1}^3}{20}$$
$$- \frac{\mathbf{u}_{j-1}^2\mathbf{u}_j}{10} - \frac{3\mathbf{u}_{j-1}\mathbf{u}_j^2}{20} - \frac{3\mathbf{u}_j^2\mathbf{u}_{j+1}}{20} - \frac{\mathbf{u}_j\mathbf{u}_{j+1}^2}{10}.$$

We can rewrite (6.7) as

$$\mathbf{u}'(t) = 6M^{-1} \left(-\frac{1}{h^2}S\mathbf{u}(t) + \lambda G(\mathbf{u}(t)) \right)$$
$$= 6M^{-1} \left(-(N+1)^2 S\mathbf{u}(t) + \lambda G(\mathbf{u}(t)) \right). \tag{6.8}$$

The results of Section 6.3.1 concerning the long-time behavior roughly carry over to the ordinary differential equation (6.8), see [LSS94], [Lar99]. In detail the following properties hold:

- There is an a-priori bound for solutions to initial values in V_h. Hence, (6.8) defines a dynamical system $S_h : \mathbb{R}_+ \times V_h \to V_h$.

- Analogously, there is a Lyapunov function $\mathcal{L} : V_h \to \mathbb{R}$, hence, S_h is a gradient system. Further on, for every $h > 0$ there is an absorbing set in V_h, i.e. S_h is dissipative.

- The bifurcation points for fixed $h = (N+1)^{-1} > 0$ are given by

$$\lambda_{h,n} = \frac{6}{h^2} \frac{1 - \cos(n\pi h)}{2 + \cos(n\pi h)}, \quad n = 1, \ldots, N,$$

cf. the linear parabolic case in (2.62). In particular, the bifurcation points converge from above to the bifurcation points of the PDE case, $\lambda_{h,n} \xrightarrow{h \to 0} \lambda_n$. Furthermore, the set $\mathbb{G}_{h,\lambda}$ of fixed points is discrete with a shape analogous to that in the PDE case.

By Theorem 6.2, the attractor $\mathcal{A}_h$ is of the same shape as above, $\mathcal{A}_h = W^u(\mathbb{G}_{h,\lambda})$. It is well-known that the attractor of the Chafee-Infante problem is upper semi-continuous, i.e. with the non-symmetric Hausdorff distance d_{NH} on $H_0^1(\Omega)$ we get

$$\lim_{h \to 0} d_{\mathrm{NH}}(\mathcal{A}_h, \mathcal{A}) = 0.$$

Furthermore, strong shadowing properties hold for this parabolic equation, see [Pil99] for details.

Lemma 5.2 implies that the support of every invariant measure is a subset of the global attractor. Recall that under proper assumptions, the AIM algorithm approximates SRB measures as defined in (1.11), i.e. invariant measures with some nice additional properties. An existence theory of SRB measures in arbitrary, high-dimensional systems is out of reach. See for instance [You02], [MY02] for an overview of the current state of research in this area.

In the ODE case it is easy to see that the ergodic measures are given as linear combinations of the Dirac measures corresponding to the fixed points of the system. See the diploma thesis [Kem02] for a detailed study of a model case. In particular, SRB measures of (6.8)—being ergodic by definition—consist of linear combinations of Dirac measures of some fixed points. Indeed we will see in the following examples that Dirac measures of the nontrivial fixed points are approximated by the subdivision algorithms.

Recall that the subdivision algorithms are formulated for discrete dynamical systems. Thus, we finally use Euler's method for time discretization and derive from (6.8) the following discrete dynamical system:

$$\mathbf{u}_{i+1} = F(\mathbf{u}_i), \quad i = 1, 2, \dots,$$

where $F : \mathbb{R}^N \to \mathbb{R}^N$ is defined by

$$F(\mathbf{u}) = \mathbf{u} + \Delta t \left(6M^{-1}(-(N+1)^2 S\mathbf{u} + \lambda G(\mathbf{u}))\right). \tag{6.9}$$

Note that we have to satisfy the stability restriction

$$\frac{\Delta t}{h^2} \le \frac{1}{2} \quad \Longleftrightarrow \quad \Delta t \le \frac{1}{2(N+1)^2}. \tag{6.10}$$

We will analyze the system given by (6.9) in the following.

6.3.3   Multiple eigenvalue 1 of the Perron-Frobenius matrix

Recall that the Perron-Frobenius Theorem A.2 for nonnegative matrices guarantees the existence of the eigenvalue 1 of the Perron-Frobenius matrix and of a corresponding nonnegative eigenvector.

We argued above that the approximated measure of the system given by (6.9) consists of a convex sum of a finite number of Dirac measures. Since every Dirac measure of a fixed point is an invariant measure of the system we expect that the Perron eigenspace of the Perron-Frobenius matrices P_k is multi-dimensional, i.e. the Perron eigenvalue 1 of P_k is geometrically multiple. In order to obtain as much as possible of the dynamics of our system, we are interested in an equally weighted convex combination of the nonnegative eigenvectors approximating the Dirac measures.

Typical eigenproblem solvers (e.g. `eigs` in MATLAB) will give us an orthonormal basis of the eigenspace corresponding to the eigenvalue 1. But in general these vectors may have negative entries although there exists a basis of nonnegative vectors corresponding to the Dirac measures. We discuss two approaches to this problem.

Nonnegative basis of Perron eigenspace: The Drazin algorithm

As already mentioned in Section 5.3.1, one can use an ansatz from the theory of generalized inverses to face the problem. Observe that the Perron eigenspace $E(P_k, 1)$ of the stochastic, nonnegative matrix P_k is given by the eigenspace $E(A, 0)$ corresponding to the eigenvalue 0 of $A := I - P_k$. Recall that the index of A as defined in 5.4 satisfies

$$\text{ind}_0(A) = \text{ind}_1(P_k) = 1$$

since P_k is stochastic. By Lemma 5.5 it follows that

$$Z_{P_k} = I - AA^D = I - AA^\# = I - (I - P_k)(I - P_k)^\# \tag{6.11}$$

is the orthogonal projection onto $E(A, 0) = E(P_k, 1)$ along $\bigoplus_{\lambda \neq 1} E(P_k, \lambda)$.

By a resolvent estimate the following theorem is shown in [HNR90]:

Theorem 6.3. *Let $P \in \mathbb{R}^{N,N}$ be a nonnegative matrix with Perron root 1 and projection Z_P as above. Then for sufficiently small $\varepsilon > 0$ the matrix*

$$C_\varepsilon = \left[(1 + \varepsilon)I - Z_P\right]^{-1} Z_P \tag{6.12}$$

is nonnegative and $\text{span} \, C_\varepsilon = \text{span} \, Z_P = E(P, 1)$.

Further on, Hartwig, Neumann and Rose give an explicit range for $\varepsilon > 0$ such that (6.12) is nonnegative. In our case this range turns out to be $\mathbb{R}_+$.

For the numerical realization of this result we need to derive the Drazin inverse numerically. There are a couple of algorithms known from numerical linear algebra, e.g. the method by Hartwig [Har81] and the shuffle algorithm by Anstreicher and Rothblum [AR87]. In our case, the index $\text{ind}_0(A) = 1$ is known. We use the following representation of the Drazin inverse for given index going back to Cline in 1968.

Theorem 6.4 ([Cli68]). *For any matrix A with index ℓ it holds that*

$$A^D = A^\ell (A^{2\ell+1})^\dagger A^\ell$$

and

$$AA^D = A^\ell (A^{2\ell})^\dagger A^\ell, \tag{6.13}$$

where $B^\dagger$ is the Moore-Penrose inverse of the matrix B.

It is well known, that the Moore-Penrose inverse $B^\dagger$ of a matrix B can be calculated by the singular value decomposition of B. This ansatz is implemented in the MATLAB algorithm pinv which we use in our numerical experiments.

Remark. *Observe that the costs for computing generalized inverses of a matrix $B \in \mathbb{R}^{K,K}$ are roughly of the order $\mathcal{O}(K^3)$ (cf. the costs for computing inverses). In addition, when using pinv, a full singular value decomposition of a non-sparse matrix is calculated. Thus, storage limitations arise. For a high number of boxes ($\approx 10^4$) the Drazin Algorithm indeed fails, as we will see below. We use the following approach as an alternative for such cases. The so-called basis rotation algorithm is already described in [Kem08]. In our computations it turns out that this algorithm works fine in higher dimensions, but is less stable than the Drazin approach.*

Basis rotation algorithm

We set up an algorithm for transferring an orthonormal basis $\{v_1, \dots, v_k\}$ of the Perron eigenspace into another orthonormal basis $\{p_1, \dots, p_k\}$ with nonnegative vectors p_i. For this, we arrange the vectors in matrices $V = \mathrm{col}(v_1, \dots, v_k)$, $P = \mathrm{col}(p_1, \dots, p_k) \in \mathbb{R}^{N,k}$. We are searching for an orthonormal matrix $O \in \mathbb{R}^{k,k}$ with

$$P = VO.$$

We construct O as a composition of Givens matrices

$$O = O_1(\alpha_1)O_2(\alpha_2) \cdots O_{k-1}(\alpha_{k-1})$$

with

$$O_\ell(\alpha_\ell) = (o_{ij})_{ij},$$
$$o_{i,j \in \{\ell, \ell+1\}} = \begin{pmatrix} \cos(\alpha) & -\sin(\alpha) \\ \sin(\alpha) & \cos(\alpha) \end{pmatrix},$$
$$o_{ij} = \delta_{ij}, \qquad i \notin \{\ell, \ell+1\} \text{ or } j \notin \{\ell, \ell+1\}.$$

By this, we rotate the basis vectors one by one into the cone of nonnegative vectors. In detail, we obtain $P = P_k$ recursively by setting

$$P_0 = V, \quad P_\ell = (p_{ij}^{(\ell)})_{ij} = P_{\ell-1}O_\ell(\alpha_\ell), \quad \ell = 1, \dots, k-1.$$

Here, the angles α_ℓ, $\ell = 1, \dots, k-1$ are given as a solution of the following minimization problem for the entries $p_{ij}^{(\ell)} = p_{ij}^{(\ell)}(\alpha_\ell)$, $i = 1, \dots, N$, $j \in \{\ell, \ell+1\}$ of P_ℓ:

$$\min\left(\sum_{\substack{i=1 \\ p_{i\ell}^{(\ell)}>0}}^{N} p_{i\ell}^{(\ell)}, - \sum_{\substack{i=1 \\ p_{i\ell}^{(\ell)}<0}}^{N} p_{i\ell}^{(\ell)} \right) + \min\left(\sum_{\substack{i=1 \\ p_{i,\ell+1}^{(\ell)}>0}}^{N} p_{i,\ell+1}^{(\ell)}, - \sum_{\substack{i=1 \\ p_{i,\ell+1}^{(\ell)}<0}}^{N} p_{i,\ell+1}^{(\ell)} \right) \overset{!}{=} \min. \qquad (6.14)$$

The minimization problem (6.14) in $\alpha_\ell \in [0, 2\pi[$ is solved by a one-dimensional root solver. We expect that each column p_i of

$$P = V\, O_1(\alpha_1)\, O_2(\alpha_2) \cdots O_{k-1}(\alpha_{k-1})$$

either is nonnegative or nonpositive and therefore a good representation of a Dirac measure in a fixed point of the system. By summing up and normalizing the columns we get a suitable new fixed point u of P with $\|u\|_1 = 1$,

$$u := \frac{v}{\|v\|_1}, \quad v = \sum_{i=1}^{k} |p_i|.$$

6.3.4 Numerical experiments for the Chafee-Infante problem

The AIM vs. the PODAIM algorithm

As a first test example we compare the results of the AIM and PODAIM algorithms for system (6.9) with the following parameters:

$$N = 6, \qquad \Delta t = 0.001, \qquad \lambda = 80. \qquad (6.15)$$

Chafee–Infante, λ = 80, N = 6, Δ t = 0.001,
AIM algorithm: 60 recursions, 107346 support boxes.

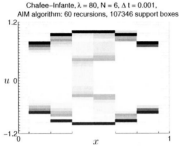
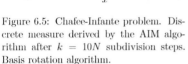

Chafee–Infante, λ = 80, N = 6, Δ t = 0.001,
2 POD modes: k = 20 recursions, 233 support boxes.

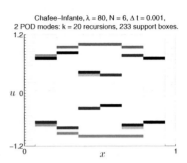

Figure 6.5: Chafee-Infante problem. Discrete measure derived by the AIM algorithm after $k = 10N$ subdivision steps. Basis rotation algorithm.

Figure 6.6: Chafee-Infante problem. Discrete measure derived by the PODAIM algorithm after $k = 10\ell$ subdivision steps in the 2-dimensional POD space. Drazin algorithm.

By Theorem 6.1, we expect 4 nontrivial fixed points of the system whose Dirac measures should be detected by the invariant measure algorithms. See Figure 6.3 for a sketch of these fixed points. Due to symmetry, they are located in a two-dimensional subspace of $\mathbb{R}^N$. Indeed, the PODAIM algorithm computes a POD space of dimension $\ell = 2$. In Figure 6.5, we see the plot of the marginal operator h_{μ_k} introduced in Section 6.1. The discrete measure μ_k is computed by the AIM algorithm with 60 subdivision steps, i.e. 10 bisections of the starting box $B_0^{(N)} = B(0, 1.2)$ in each space dimension. The four nontrivial fixed points are well detected. Note that the basis rotation algorithm is used since the number of support boxes becomes very large.

In Figure 6.6 we see the result of the PODAIM algorithm in dimension $\ell = 2$ with starting box $B_0^{(\ell)} = B(0, 2.5)$. The discrete measure is illustrated after $k = 20$ subdivision steps, also corresponding to 10 bisections in each space dimension. The Dirac measures of the nontrivial fixed points are obviously even better approximated than in the AIM algorithm. This is due to the fact that the linear two-dimensional POD subspace approximates the subspace containing the fixed points very accurate. Nevertheless, there are still some support boxes away from the fixed points as indicated by almost white boxes covering the grid in the marginal representation. This can also be observed in the following figures.

The most remarkable result of this example is the number of boxes building the support of the discrete measures. In the AIM algorithm—despite the simple form of the approximated invariant measure—the support of the discrete measure consists of more than 100 000 boxes while in the PODAIM algorithm, only 233 boxes are needed to produce a comparable result.

In Chapter 4 we have introduced the Prohorov metric as a suitable distance notion for discrete measures computed by the subdivision algorithms. We use the discretization of the Prohorov metric given by Algorithm 4.3 to derive distances between the discrete measures in different subdivision steps. The results are plotted in Figures 6.7 and 6.8. We analyze two approximation processes: For the blue curves, the resulting POD measure $\mu_{20}^{(\text{pod})}$ after 10 bisections in each component is used as a reference measure and the distance between the discrete AIM measure μ_k and this reference measure is computed in the discretized

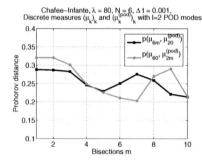

Figure 6.7: Chafee-Infante problem. Prohorov distance of the discrete measures of the AIM and PODAIM algorithm for the first $k = 10N$ and $k = 10\ell$ recursion steps, respectively.

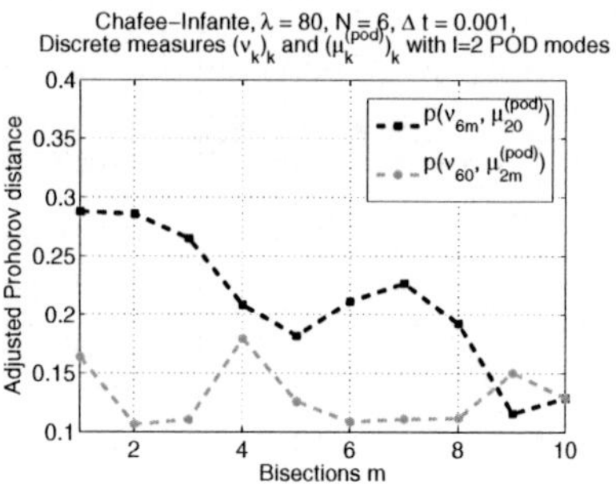

Figure 6.8: Chafee-Infante problem. Prohorov distance for the same discrete measures as in Figure 6.7. Adjusted support box collection of the AIM measure.

Prohorov metric for the first 10 bisection steps, i.e. $k = 6m$, $1 \le m \le 10$. For the red curves, we use the resulting AIM measure μ_{60} after 10 bisections in each component as a reference measure and compute the distance between μ_{60} and $\mu_k^{(\text{pod})}$ for the first 10 bisection steps as described above. In this way, the approximation process of the reduced system measures is illustrated.

In Figure 6.7, we see that the Prohorov metric does not give very convincing results. Although by and large the distance decreases, still after 10 bisections the remainder is quite large compared to the distance in the first steps. We suggest two explanations for this behavior: First, observe that the projection of the starting box $B_0^{(N)}$ into the reduced space $\mathbb{R}^2$, given by $W_1 W_1^T(B_0^{(N)})$, has a large diameter due to the incompatibility of the maximum norm defining the starting box and the Euclidean norm related to the orthogonal POD matrix $W_1 \in \mathbb{R}^{6,2}$. Hence, the projection of support boxes of the AIM measure may be located far away from the support of the reduced-space measure just by numerical artefacts. Figure 6.8 supports these considerations. For the plots in this figure, the support boxes with projected centers c of norm larger than 2, i.e. $\|W_1 W_1^T c\|_\infty > 2$, have been taken out of the support of μ_k to obtain the value of the measure ν_k. By this, the decrease of the resulting Prohorov distance is somewhat better than that of the original measures.

The second reason for the slow decrease is given by the bad approximation process of the AIM algorithm in this case. As we have seen in the marginal representation, an invariant measure of the original system is not approximated very well even after $k = 60$ recursion steps. This is also supported by the curves of Figure 6.8. Even the adjusted measure ν_{60} is not approached closely by the reduced-system measures during the refinement process.

Different reduced space dimensions in the PODAIM algorithm

In Figures 6.9, 6.10 and 6.11, we show the influence of the POD space dimension on the PODAIM algorithm. Recall that we derived a bound for the distance between discrete measures in reduced-space and original state space in a special setting in Corollary 5.14.

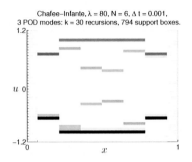

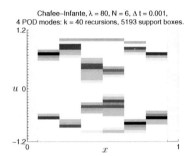

Figure 6.9: Chafee-Infante in different POD space dimensions. Discrete measures after $k = 10\ell$ recursion steps of the PODAIM algorithm for $\ell = 3$ and $\ell = 4$.

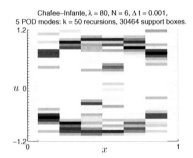

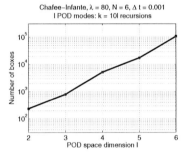

Figure 6.10: Chafee-Infante in different POD space dimensions. Discrete measure after $k = 10\ell$ recursions for $\ell = 5$ by use of the basis rotation algorithm.

Figure 6.11: Chafee-Infante in different POD space dimensions. Number of boxes after $k = 10\ell$ recursion steps.

The bound (5.46) depends on the POD space dimension ℓ. In the following, we analyze the dependence of the reduced-space discrete measures on the dimension ℓ in the Chafee-Infante problem.

We choose the same system parameters and stepsizes as in our first numerical computation, $(\lambda, N, \Delta t) = (80, 6, 10^{-3})$, and compare discrete measures after $k = 10\ell$ recursion steps. We raise the number of POD modes and, in addition to the discrete measure for $\ell = 2$ plotted in Figure 6.6, get discrete measures for $\ell = 3$ and $\ell = 4$ plotted in Figure 6.9. In Figure 6.10, the discrete measure for the case $\ell = 5$ is plotted. Due to storage limitations we fall again short of deriving a result for $k = 10\ell = 50$ with the Drazin algorithm and use the basis rotation algorithm instead.

Our first observation is that the number of boxes and hence the overall computational effort after $k = 10\ell$ recursion steps increases exponentially with the state space dimension ℓ, see Figure 6.11.

Secondly, these computational efforts do not lead to better approximation results for an invariant measure of the original system: Comparing the marginal representations in Figures 6.6, 6.9, 6.10 and 6.5, we see that we lose accuracy in the representation of the Dirac measures with each additional POD space dimension.

The PODAIM algorithm in high dimensions

We give the prospects for the chances and limitations of the PODAIM algorithm by looking at higher original state space dimensions $N \in \mathbb{N}$. Based on the first computation in the POD space for $N = 6$ with the resulting discrete measure plotted in Figure 6.6, we compute discrete measures in the POD spaces for the following state space dimensions:

$$N = \{12, 24, 48, 96\}.$$

In all computations, we fix the system parameter $\lambda = 80$ and the number of POD vectors given by $\ell = 2$. Due to the stability condition (6.10), the step size Δt is quartered when the space dimension N is doubled. We work with fixed integration time 0.001 for the AIM subdivision process in order to get comparable results.

If we look at the marginal-like representations of the discrete measures plotted in Figures 6.12 and 6.13, we see that the fixed points are well detected in all cases. Fortunately, the number of support boxes remains of the same order during the spatial refinement process.

Nevertheless, the computational effort increases significantly with the original state space dimension as we see in Table 6.2. This is due to the fact that although the reduced-space system function F_{red} operates on a low-dimensional space, an evaluation of F_{red} needs an evaluation of F, since $F_{\text{red}}(\alpha) = W^T F(W\alpha)$.

Space dimension N	6	12	24	48	96
CPU time in s	3.88	8.71	215.58	386.47	2817.60

Table 6.2: Chafee-Infante in increasing space dimensions. CPU time needed for the evaluation of the PODAIM measure after $k = 10\ell$ recursion steps on a standard PC.

The PODADAPT algorithm

At the end of this section, we give the prospects for the PODADAPT algorithm described in Section 3.2. As a test example we use the same system as in the first example, i.e. the parabolic system given by (6.15).

We start the PODADAPT algorithm with the initial data

$$(\mathcal{B}_0, \mu_0, l_0, W_0) = \left(\{B_0^{(N)}\}, \mu_0, 6, I_6 \right)$$

Here, $B_0^{(N)}$ is the starting box of the first example and μ_0 is the discrete measure on $\mathcal{B}_0$ with constant density on $B_0^{(N)} = \operatorname{supp}\mu_0$. Observe that the initial POD space is simply given by the original state space $\mathbb{R}^N$ with standard basis. Hence, the algorithm starts exactly as the AIM algorithm for this problem.

The first computation of a new POD basis is performed after $k = 12$ recursion steps, or in other words after bisecting two times in each coordinate. As proposed in Algorithm

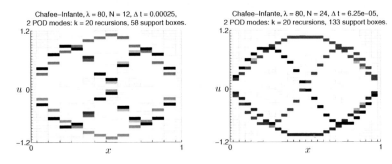

Figure 6.12: Chafee-Infante in increasing space dimensions. PODAIM algorithm for different discretization step sizes. Discrete measures after $k = 10\ell = 20$ recursions, $(N, \Delta t) = (12, \frac{1}{4} \cdot 10^{-3})$ and $(N, \Delta t) = (24, \frac{1}{16} \cdot 10^{-3})$.

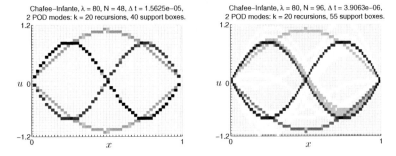

Figure 6.13: Chafee-Infante in increasing space dimensions, continued. PODAIM algorithm for different discretization step sizes. Discrete measures after $k = 10\ell = 20$ recursions, $(N, \Delta t) = (48, \frac{1}{64} \cdot 10^{-3})$ and $(N, \Delta t) = (96, \frac{1}{256} \cdot 10^{-3})$.

3.3, the box collection is transformed and a new discrete measure is derived. In the new POD space, ℓ_k subdivision steps are performed before the POD space is being adapted the next time. The resulting discrete measure in Figure 6.14 was derived after 20 bisections in each coordinate, i.e. 18 adaptations of the POD space. This corresponds to the total amount of 61 subdivision steps since this is the sum of POD dimensions occurring in the adaptation steps of the algorithm, see Table 6.3.

As noted in Section 3.2, there are 3 parameters in the adaptation step of the algorithm.

bisections per component	1	2	3	4	5	6	7	8	9	10	11	12	13	14	15	16	17	18	19	20
POD dimension	6	6	6	3	2	3	3	3	3	3	3	3	3	2	2	2	2	2	2	2

Table 6.3: PODADAPT: Computed POD space dimensions in the adaptation steps.

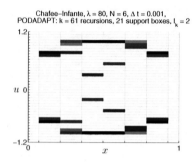

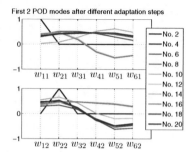

Figure 6.14: Chafee-Infante with the PO-DADAPT algorithm. Discrete measure after $k = 61$ subdivision steps corresponding to 20 bisections in each component.

Figure 6.15: Chafee-Infante with the PO-DADAPT algorithm. Evolution of the first two POD modes during the adaptation steps.

First, the number ℓ_k of POD modes was chosen according to

$$\ell_k = \min\left\{6, \min\{\ell > 0 : \frac{\sigma_\ell}{\sigma_1} < 0.05\}\right\}$$

where $\sigma_1 \geq \ldots \geq \sigma_6 \geq 0$ are the singular values of the matrix of snapshots. The other two parameters that define the collection of snapshots are given by $(m, T) = (10\,000, 10)$. Recall that $m \in \mathbb{N}$ denotes the number and $T \in \mathbb{N}$ the length of the trajectories collected as snapshots. Observe that this choice of parameters was rejected in the POD approximation step of the Lorenz system. Here, the choice leads to satisfying results as we see in Figure 6.14. The information of all nontrivial fixed points is included in the resulting discrete measure.

In Figure 6.15 we see how the POD adaptation evolves. After starting with the canonical basis, the first adapted POD vectors represent the stable and the unstable fixed points in this order. After a few further adaptation steps, the order of the POD modes has changed. At the end, when the number of POD modes decreases to $\ell = 2$, the order of the beginning is reached again. Despite this dynamic evolution of the POD modes, no information is lost if we compare the resulting discrete measure with the PODAIM measure of Figure 6.6. Only the shape of the unstable fixed points is slightly perturbed.

Nevertheless, we note that in this example the use of the adaptation is not necessary to get good approximation results of the discrete measures. The additional effort of the adaptation does not give additional benefit and thus it suffices to take the PODAIM algorithm for this case. It is part of future work to find applications where the PODADAPT algorithm has computational advantages over the algorithm PODAIM in fixed POD space.

Appendix A

Appendix

A.1 Nonnegative matrices

There is plenty of literature concerning the theory of nonnegative matrices, see e.g. [BP79], [Min88], [HJ85]. In this thesis we use the following definitions and facts.

Definition A.1. A square matrix $A = (a_{ij})_{ij} \in \mathbb{R}^{n,n}$ is called *nonnegative*, $A \geq 0$, if all entries are nonnegative

$$a_{ij} \geq 0.$$

The matrix is called *positive*, $A > 0$, if all entries are positive. Further on a nonnegative matrix $A \in \mathbb{R}^{n,n}$, $n \geq 2$, is called *reducible* if it is cogredient to a matrix

$$E = \begin{pmatrix} B & 0 \\ C & D \end{pmatrix}$$

where B, D are square matrices. For $n = 1$, the only reducible matrix is given by $A = 0$. If $A \in \mathbb{R}^{n,n}$ is not reducible, the matrix is called *irreducible*.

Theorem A.2 (Perron-Frobenius). *Let $A \in \mathbb{R}^{n,n}$ be a nonnegative matrix. Then the following holds:*

- *The spectral radius $\rho(A)$ is an eigenvalue of A.*

- *A has a nonnegative eigenvector $v \geq 0$ corresponding to $\rho(A)$:*

$$Av = \rho(A)v. \tag{A.1}$$

- *A^T also has a nonnegative eigenvector corresponding to $\rho(A)$.*

- *The spectral radius is bounded by the row sums of A: With*

$$s = \min_{j=1,\dots,n} \sum_{i=1}^{n} a_{ij}, \quad S = \max_{j=1,\dots,n} \sum_{i=1}^{n} a_{ij}$$

the following inequality holds:

$$s \leq \rho(A) \leq S. \tag{A.2}$$

123

Inequality (A.2) *holds with equality if and only if* $s = S$.

If A is irreducible then $\rho(A)$ is simple, every other eigenvalue on the spectral circle

$$\partial\sigma(A) = \{\lambda \in \mathbb{C} : |\lambda| = \rho(A)\}$$

is also simple and A has a positive eigenvector corresponding to $\rho(A)$.

If A is primitive, *i.e. $A^m > 0$ for an $m \in \mathbb{N}$, the eigenvalue $\rho(A)$ is simple and the only eigenvalue on the spectral circle.*

Proof. see [BP79]. $\qquad\qquad\qquad\qquad\qquad\qquad\qquad\qquad\qquad\qquad\qquad\qquad\qquad$ □

Remark. *Observe that equation* (A.2) *also holds for column sums by considering A^T instead of A.*

A.2   Perturbation theory for matrices

We collect some properties of the singular value decomposition of a matrix $A \in \mathbb{C}^{m,n}$. The singular value decomposition of A is given by

$$W^H A V = \begin{pmatrix} \Sigma \\ 0 \end{pmatrix} \tag{A.3}$$

where $W \in \mathbb{C}^{m,m}$ and $V \in \mathbb{C}^{n,n}$ are unitary matrices and $\Sigma = \mathrm{diag}(\sigma_1, \ldots, \sigma_n)$ with $\sigma_1 \geq \ldots \geq \sigma_n \geq 0$. It is easy to see, that this decomposition exists for every $A \in \mathbb{C}^{m,n}$. It follows immediately from (A.3) that the columns of W, also called *left singular vectors*, are the eigenvectors of AA^H and the columns of V, also called *(right) singular vectors* are the eigenvectors of $A^H A$.

The singular values can be represented as eigenvalues of a larger matrix via the following theorem

Theorem A.3 (Jordan-Wielandt, see [GvL96]). *Let $A \in \mathbb{C}^{m,n}$, $m \geq n$, be given with singular value decomposition*

$$W^H A V = \begin{pmatrix} \Sigma \\ 0 \end{pmatrix}$$

where $\Sigma = \mathrm{diag}(\sigma_1, \ldots, \sigma_n)$ and $W = \mathrm{col}(w_1, \ldots, w_m) \in \mathbb{C}^{m,m}$ and $V = \mathrm{col}(v_1, \ldots, v_n) \in \mathbb{C}^{n,n}$ unitary as above. Then the eigenvalues of

$$C = \begin{pmatrix} 0 & A^H \\ A & 0 \end{pmatrix} \in \mathbb{C}^{m+n,m+n}$$

are given by $\pm\sigma_1, \ldots, \pm\sigma_n$ with corresponding eigenvectors $\begin{pmatrix} w_i \\ \pm v_i \end{pmatrix}$ and $m - n$ zero eigenvalues with corresponding eigenvectors $\begin{pmatrix} w_i \\ 0 \end{pmatrix}$, $i = n + 1, \ldots, m$.

Proof. A simple computation reveals

$$Q^T \begin{pmatrix} 0 & A^H \\ A & 0 \end{pmatrix} Q = \mathrm{diag}(\sigma_1, \ldots, \sigma_n, -\sigma_1, \ldots, -\sigma_n, 0, \ldots, 0) \in \mathbb{C}^{m+n,m+n}$$

if we define the unitary(!) matrix $Q \in \mathbb{C}^{m+n,m+n}$ by

$$Q = \frac{1}{\sqrt{2}} \begin{pmatrix} V & V & 0 \\ W_1 & -W_1 & \sqrt{2}W_2 \end{pmatrix}, \quad \text{where } W = \begin{pmatrix} W_1 & W_2 \end{pmatrix}, W_1 \in \mathbb{C}^{m,n}, W_2 \in \mathbb{C}^{m,m-n}.$$

□

Theorem A.4 (Mirsky). *Let A, E be Hermitian matrices with eigenvalues $\alpha_1 \geq \ldots \geq \alpha_n$ of A and $\hat{\alpha}_1 \geq \ldots \geq \hat{\alpha}_n$ of A + E. Then for every unitarily invariant norm $\|\cdot\|$ it follows*

$$\|\operatorname{diag}(\alpha_i - \hat{\alpha}_i)\| \leq \|E\|.$$

Proof. see [Ste90].   □

Remark. *The matrix norm $\|A\|_2$ induced by the Euclidean norm and the Frobenius norm $\|A\|_F = \left(\sum_{i,j=1}^n a_{ij}^2\right)^{1/2}$ are examples for unitarily invariant norms.*

We use the following interlacing property of eigenvalues of symmetric (sub-)matrices to derive a result for singular values.

Theorem A.5. *Let $A \in \mathbb{R}^{n,n}$ be symmetric and $A_r \in \mathbb{R}^{r,r}$, $r \leq n$ the r-by-r principle sub-matrix of A. Then the following holds for the ordered eigenvalues $\lambda_i(A_r)$, $i = 1, \ldots, r$ of A_r, $r = 1, \ldots, n$:*

$$\lambda_{r+1}(A_{r+1}) \leq \lambda_r(A_r) \leq \lambda_r(A_{r+1}) \leq \ldots, \leq \lambda_2(A_{r+1}) \leq \lambda_1(A_r) \leq \lambda_1(A_{r+1})$$

with $r = 1, \ldots, r - 1$.

Proof. see [GvL96].   □

Corollary A.6. *Let $A = A_n = \operatorname{col}(a_1, \ldots, a_n) \in \mathbb{R}^{m,n}$, $m \geq n \in \mathbb{N}$ and let $\sigma_i(A_r)$, $i = 1, \ldots, r$, $r = 1, \ldots, n$ be the singular values of the submatrix $A_r = \operatorname{col}(a_1, \ldots, a_r)$. Then the following interlacing property holds:*

$$\sigma_1(A_{r+1}) \geq \sigma_1(A_r) \geq \sigma_2(A_{r+1}) \geq \sigma_2(A_r) \geq \ldots \geq \sigma_r(A_{r+1}) \geq \sigma_r(A_r) \geq \sigma_{r+1}(A_{r+1})$$

for every $r = 1, \ldots, n - 1$.

Proof. This follows immediately from Theorem A.5 if we consider $A_r^T A_r \in \mathbb{R}^{r,r}$ which is the r-by-r submatrix of $A^T A \in \mathbb{R}^{n,n}$ with nonnegative eigenvalues $\sigma_i^2(A_r)$, $i = 1, \ldots, r$.   □

Theorem A.7. *Let $A \in \mathbb{C}^{m,\ell}$, $B \in \mathbb{C}^{\ell,n}$ be given and denote the singular values of A, B and $C = AB \in \mathbb{C}^{m,n}$ by $\{\sigma_i(A)\}_{i=1}^{\min(m,l)}$, $\{\sigma_i(B)\}_{i=1}^{\min(l,n)}$, $\{\sigma_i(C)\}_{i=1}^{\min(m,n)}$. Then we get the following estimate for $\sigma_i(C)$:*

$$\sigma_i(A)\sigma_{\min(l,n)}(B) \leq \sigma_i(C) \leq \sigma_i(A)\sigma_1(B), \quad \text{for all } i = 1, \ldots, \min(m,n,l).$$

Proof. see [Ste90], Theorem I.4.5 and Exercise I.4.6.   □

Theorem A.8. *Let $A, B \in \mathbb{C}^{m,n}$, $m \geq n$. Then for all $i, j = 1, \ldots, n$ such that $i + j \leq n + 1$,*

$$\sigma_{i+j-1}(A + B) \leq \sigma_i(A) + \sigma_j(B),$$
$$\sigma_{i+j-1}(A) + \sigma_j(B) \leq \sigma_i(A + B).$$

Proof. see [HJ94]. □

The following approach to the concept of canonical angles is also described in [Ste90].

Theorem A.9. *Let* $W \in \mathbb{C}^{n,n}$ *be unitary with*

$$W = \begin{pmatrix} \overset{\ell}{W_{11}} & \overset{n-\ell}{W_{12}} \\ W_{21} & W_{22} \end{pmatrix} \begin{matrix} \ell \\ n-\ell \end{matrix}$$

Then there are unitary matrices $U = \begin{pmatrix} U_{11} & 0 \\ 0 & U_{22} \end{pmatrix}$, $V = \begin{pmatrix} V_{11} & 0 \\ 0 & V_{22} \end{pmatrix} \in \mathbb{C}^{n,n}$ *with*

$$U^H W V = \begin{pmatrix} \overset{\ell}{\Gamma} & \overset{\ell}{-\Sigma} & \overset{n-2\ell}{0} \\ \Sigma & \Gamma & 0 \\ 0 & 0 & I_{n-2\ell} \end{pmatrix} \begin{matrix} \ell \\ \ell \\ n-2\ell \end{matrix}$$

where $\Gamma = \mathrm{diag}(\gamma_1, \dots, \gamma_\ell)$, $\Sigma = \mathrm{diag}(\sigma_1, \dots, \sigma_\ell)$, $\gamma_i, \sigma_i \geq 0, \gamma_i^2 + \sigma_i^2 = 1$, $i = 1, \dots, \ell$.

Proof. see [Ste90]. □

Theorem A.10. *Let* $X_1, Y_1 \in \mathbb{C}^{n,\ell}$ *be column orthonormal:* $X_1^H X_1 = Y_1^H Y_1 = I_\ell$. *Then it holds*

- *for* $2\ell \leq n$: *There are unitary* $Q \in \mathbb{C}^{n,n}, U_{11}, V_{11} \in \mathbb{C}^{\ell,\ell}$ *with*

$$QX_1 U_{11} = \begin{pmatrix} I_\ell \\ 0 \\ 0 \end{pmatrix}, \quad QY_1 V_{11} = \begin{pmatrix} \Gamma \\ \Sigma \\ 0 \end{pmatrix} \tag{A.4}$$

  and $\Gamma = \mathrm{diag}(\gamma_1, \dots, \gamma_\ell)$, $\Sigma = \mathrm{diag}(\sigma_1, \dots, \sigma_\ell)$, $0 \leq \gamma_1 \leq \dots \leq \gamma_\ell$, $\sigma_1 \geq \dots \geq \sigma_\ell \geq 0$, $\sigma_i^2 + \gamma_i^2 = 1$, $i = 1, \dots, \ell$.

- *for* $2\ell > n$: *There are unitary* $Q \in \mathbb{C}^{n,n}, U_{11}, V_{11} \in \mathbb{C}^{\ell,\ell}$ *with*

$$QX_1 U_{11} = \begin{pmatrix} I_{n-\ell} & 0 \\ 0 & I_{2\ell-n} \\ 0 & 0 \end{pmatrix}, \quad QY_1 V_{11} = \begin{pmatrix} \Gamma & 0 \\ 0 & I_{2\ell-n} \\ \Sigma & 0 \end{pmatrix} \tag{A.5}$$

  and $\Gamma = \mathrm{diag}(\gamma_1, \dots, \gamma_{n-\ell})$, $\Sigma = \mathrm{diag}(\sigma_1, \dots, \sigma_{n-\ell})$, γ_i, σ_i *as above.*

Proof. see [Ste90]. □

This theorem allows us to define canonical angles between subspaces.

Definition A.11. *For* $X = \mathrm{col}(x_1, \dots, x_\ell)$ *and* $Y = \mathrm{col}(y_1, \dots, y_\ell)$ *with orthonormal columns we define the* canonical angle *between the subspaces* $R(X)$ *and* $R(Y)$ *given by* X *and* Y *as the matrix*

$$\angle(X, Y) := \sin^{-1} \Sigma \tag{A.6}$$

where $\Sigma \in \mathbb{C}^{\ell,\ell}$ *and* $\Sigma \in \mathbb{C}^{n-\ell,n-\ell}$ *are the diagonal matrices as defined in* (A.4) *and* (A.5), *respectively.*

Remark. *Observe that*

- $\Sigma = 0$ *for* $\mathcal{X} = \mathcal{Y}$,

- $\sigma_i \in [0,1]$,

- $\Sigma = I$ *if* $\mathcal{X}$ *and* $\mathcal{Y}$ *are orthogonal.*

- *The right multiplication with a unitary matrix does not change the range. Therefore* $\angle(\hat{X}, \hat{Y}) = \angle(X,Y)$ *for* $\hat{X} := XU$, $\hat{Y} := YV$, $U, V \in \mathbb{C}^{\ell,\ell}$ *unitary.*

Now, let the singular value decompositions of A and $A + E \in \mathbb{C}^{m,n}$, $m \geq n$ be given by

$$\begin{pmatrix} W_1 & W_2 & W_3 \end{pmatrix}^H A \begin{pmatrix} V_1 & V_2 \end{pmatrix} = \begin{pmatrix} \Sigma_1 & 0 \\ 0 & \Sigma_2 \\ 0 & 0 \end{pmatrix}$$

$$\begin{pmatrix} \tilde{W}_1 & \tilde{W}_2 & \tilde{W}_3 \end{pmatrix}^H (A + E) \begin{pmatrix} \tilde{V}_1 & \tilde{V}_2 \end{pmatrix} = \begin{pmatrix} \tilde{\Sigma}_1 & 0 \\ 0 & \tilde{\Sigma}_2 \\ 0 & 0 \end{pmatrix}$$

where $\Sigma_1, \tilde{\Sigma}_1 \in \mathbb{R}^{\ell,\ell}$, $\Sigma_2, \tilde{\Sigma}_2 \in \mathbb{C}^{n-\ell,n-\ell}$, $W_1, \tilde{W}_1 \in \mathbb{C}^{m,\ell}$, $W_2, \tilde{W}_2 \in \mathbb{C}^{m,n-\ell}$, $W_3, \tilde{W}_3 \in \mathbb{C}^{m,m-n}$, $V_1, \tilde{V}_1 \in \mathbb{C}^{n,\ell}$ and $V_2, \tilde{V}_2 \in \mathbb{C}^{n,n-\ell}$.

Theorem A.12. *Let* $\alpha, \delta > 0$ *be given with*

$$\min \sigma(\tilde{\Sigma}_1) \geq \alpha + \delta \quad and \quad \max \sigma(\Sigma_2) \leq \alpha$$

Then we have

$$\max(\| \sin \angle(W_1, \tilde{W}_1)\|_2, \| \sin \angle(V_1, \tilde{V}_1)\|_2) \leq \frac{\max\{\|R\|_2, \|S\|_2\}}{\delta},$$

where R *and* S *are the residuals*

$$R := A\tilde{V}_1 - \tilde{W}_1 \tilde{\Sigma}_1 \quad and \quad S := A^H \tilde{W}_1 - \tilde{V}_1 \tilde{\Sigma}_1.$$

Proof. see [Ste90].   □

Combined with the Theorem of Mirsky, we get an estimate of the canonical angle of singular spaces in terms of spectral gaps of $A \in \mathbb{R}^{m,n}$:

Corollary A.13. *With the notation as above let the singular values be arranged in a descending order in* Σ_1, Σ_2. *Denote by* γ *the spectral gap between* Σ_1 *and* Σ_2:

$$\gamma := \min\{\sigma_1 - \sigma_2 : \sigma_1 \in \sigma(\Sigma_1), \sigma_2 \in \sigma(\Sigma_2)\} = \min \sigma(\Sigma_1) - \max \sigma(\Sigma_2).$$

Then it holds for $\gamma > \|E\|_2$:

$$\max(\| \sin \angle(W_1, \tilde{W}_1)\|_2, \| \sin \angle(V_1, \tilde{V}_1)\|_2) \leq \frac{\|E\|_2}{\gamma - \|E\|_2}.$$

Proof. Using Theorem A.12 let $\alpha = \max \sigma(\Sigma_2)$ or $\alpha = \varepsilon \ll 1$ if $\max \sigma(\Sigma_2) = 0$. In the last case a limit process $\varepsilon \to 0$ at the end gives the statement.

Let $\sigma_1 = \min \sigma(\Sigma_1)$. Then we get by the Theorem of Mirsky:

$$|\min \sigma(\tilde{\Sigma}_1) - \sigma_1| \le \|E\|_2$$

and by that

$$\min \sigma(\tilde{\Sigma}) \ge \sigma_1 - \|E\|_2 = \alpha + \gamma - \|E\|_2.$$

Theorem A.12 with $\delta = \gamma - \|E\|_2$ implies the statement by use of $\|R\|_2, \|S\|_2 \le \|E\|_2$. $\square$

A.3   Aspects of linear functional analysis

Let H be a complex separable Hilbert space with scalar product $\langle \cdot, \cdot \rangle$. We consider the following sets of linear operators on H:

- the set of linear continuous operators:

$$L(H) = \{T : H \to H : H \text{ is linear and continuous}\},$$

- the set of compact operators:

$$K(H) = \{T \in L(H) : \overline{T(B_1(0))} \text{ is compact in } H\}$$

- the set of compact self adjoint operators:

$$S(H) = \{T \in K(H) : \langle\, Tx, y \,\rangle = \langle\, x, Ty \,\rangle \text{ for all } x, y \in H\}.$$

Theorem A.14 (Spectral theorem for selfadjoint compact operators). *Let $T \in S(H)$ be given with $T \ne 0$. Then the spectrum $\sigma(T)$ satisfies the following properties:*

- $\sigma(T) \setminus \{0\}$ *is given by countable (finite or infinite) many complex numbers with no other limit point than 0.*

- *There exists an orthonormal system $(e_k)_{k \in N}$ and a sequence $(\lambda_k)_{k \in N}$, $\lambda_k \ne 0$, $N \subset \mathbb{N}$, such that*

$$Te_k = \lambda_k e_k, \quad k \in N, \quad \sigma(T) \setminus \{0\} = \{\lambda_k : k \in \mathbb{N}\}.$$

- *The index for all eigenvalues is equal to 1:*

$$\mathrm{ind}_{\lambda_k}(T) := \max\{n \in \mathbb{N} : N(\lambda_k I - T)^{n-1} \ne N(\lambda_k I - T)^n\} = 1, \quad \text{for all } k \in N.$$

- *We get the following decomposition*

$$H = N(T) \perp \overline{\mathrm{span}\{e_k\}_k},$$

where $A \perp B$ is the orthogonal sum of the subspaces $A, B \subset H$.

- *In particular we get for every $x \in H$:*

$$Tx = \sum_{k \in N} \lambda_k \langle x, e_k \rangle.$$

Proof. see [Alt99]. □

Special linear operators are given by projections:

Definition A.15. Let V be a subspace of the Hilbert space H. $P \in L(H)$ is called a *projection onto* V, iff

$$P^2 = P, \qquad R(P) = V.$$

The *orthogonal projection onto* V is defined by $P : H \to V$ with

$$\|v - Pv\| \leq \|v - Pw\|, \quad \text{for all } v, w \in H.$$

A.4 Topics from measure and integration theory

We recall some basic definitions and constructions concerning measures and integration in general measure spaces. Most of the following is standard theory. For proofs see [Fol99], [Bau92], [Els05] and [Bil99], respectively.

For later references we first define some function spaces.

Definition A.16. Let $\Omega \subset \mathbb{R}^n$ be open. For $m \in [0, \infty]$ we denote by $C^m(\Omega)$ the vector space of functions with continuous derivatives $\partial^\alpha f$ up to order $|\alpha| \leq m$ on Ω. In particular, let $C(X) = C^0(X)$ be the space of continuous functions on X.

Denote the space of $C^m(\Omega)$-functions with compact support by $C_c^m(\Omega)$, the set of bounded continuous functions by $C_b^m(\Omega)$ and, for $\Omega = R^n$, the set of continuous functions vanishing at infinity by $C_0^m(\mathbb{R}^n)$.

For $\Omega = \mathbb{R}^n$ we often omit the notion of the space and write $C^m = C^m(\mathbb{R}^n)$, $C_0^m = C_0^m(\mathbb{R}^n)$ and so on.

Measure spaces

Definition A.17. Let X be a nonempty set. $\mathcal{A} \subset \mathcal{P}(X)$ is called σ-*algebra* iff for every $E \in \mathcal{A}$ we have $E^c \in \mathcal{A}$ and for every sequence of sets $E_i \in \mathcal{A}$, $i \in \mathbb{N}$, we have $\bigcup_{i=1}^{\infty} E_i \in \mathcal{A}$. For a σ-algebra $\mathcal{A}$ the touple $(X, \mathcal{A})$ is called *measurable space*, $A \in \mathcal{A}$ *measurable set*.

For $\mathcal{E} \subset \mathcal{P}(X)$ we call $\mathcal{A}(\mathcal{E})$ the σ-algebra *generated by* $\mathcal{E}$, iff

$$\mathcal{A}(\mathcal{E}) := \bigcap \{\mathcal{A} \subset \mathcal{P}(X) : \mathcal{A} \text{ is a } \sigma\text{-algebra of } X, \mathcal{E} \subset \mathcal{A}\}.$$

We call $\mathcal{B}(X)$ the Borel σ-algebra, iff $\mathcal{B}(X)$ is generated by all open sets of a topological space X.

Remark. *Equivalently, $\mathcal{B}(X)$ is generated by all closed sets. In particular $\mathcal{B}(\mathbb{R})$ is generated by all open, closed or half-open intervals.*

Definition A.18. Let $\{(X_\alpha, \mathcal{A}_\alpha)\}_{\alpha \in A}$ be an indexed collection of nonempty measurable spaces, $X = \prod_{\alpha \in A} X_\alpha$ and $\pi_\alpha : X \to X_\alpha$ the coordinate maps. Then the *product σ-algebra* $\otimes_{\alpha \in A} \mathcal{A}_\alpha$ is defined as the σ-algebra generated by

$$\{\pi_\alpha^{-1}(E_\alpha) : E_\alpha \in \mathcal{A}_\alpha, \alpha \in A\}.$$

Proposition A.19. *For separable metric spaces $X_1, \ldots, X_n$ and $X = \prod_{i=1}^n$ equipped with the product metric it holds*

$$\otimes_{i=1}^n \mathcal{B}(X_i) = \mathcal{B}(X).$$

In particular $\mathcal{B}(\mathbb{R}^n) = \otimes_{i=1}^n \mathcal{B}(\mathbb{R})$.

Definition A.20. Let $(X, \mathcal{A})$ be a measurable space. Then a *measure* on $\mathcal{A}$ (sometimes called measure on X if the choice of $\mathcal{A}$ is clear) is a function $\mu : \mathcal{A} \to [0, \infty]$ with $\mu(\emptyset) = 0$ and

$$\mu\left(\bigcup_{i=1}^\infty E_j\right) = \sum_{i=1}^\infty \mu(E_j) \quad \text{for all sequences of disjoint sets } \{E_i\}_{i=1}^\infty.$$

With μ as above $(X, \mathcal{A}, \mu)$ is called a *measure space*.

A measure is called *finite*, iff $\mu(X) < \infty$, *probability measure* if $\mu(X) = 1$ and *σ-finite*, iff there exists a sequence of sets $E_i \subset X$, $i \in \mathbb{N}$, with $\mu(E_i) < \infty$ and $\cup_{i=1}^\infty E_i = X$.

Two important measures are given by the following proposition

Proposition A.21. *Let $x_0 \in X$ and $(X, \mathcal{A})$ be an arbitrary measurable space. Define $\delta_{x_0} = \delta_{x_0} : \mathcal{A} \to [0, 1]$ by*

$$\delta_{x_0}(A) = 1, \ \text{if } x_0 \in A, \ \delta_{x_0}(A) = 0 \ \text{otherwise}, \quad \text{for all } A \in \mathcal{A}. \tag{A.7}$$

Then δ_{x_0} is a probability measure on X, called the Dirac measure *at x_0.*

The Lebesgue measure *on the real line is uniquely defined by the measure $\lambda : \mathcal{L} \to [0, \infty]$ with*

$$\lambda((a, b)) = b - a.$$

Remark. *If not stated otherwise we define $\delta_{x_0} : \mathcal{B} \to [0, 1]$ on the Borel σ-algebra $\mathcal{B}(X)$.*

For details about the construction of the Lebesgue measure via outer measures and premeasures see [Fol99], Sections 1.4, 1.5. By construction the Lebesgue measurable sets $\mathcal{L}$ are unions of Borel sets and Lebesgue null sets. We identify the restriction of the Lebesgue measure to the Borel sets with λ_n.

Definition A.22. Let $(X_1, \mathcal{A}_1, \mu)$, $(X_2, \mathcal{A}_2, \nu)$ be measure spaces. We call a set $A \times B$ with $A \in \mathcal{A}_1$, $B \in \mathcal{A}_1$ a *rectangle*. Define the premeasure π on the product σ-algebra $\mathcal{A}_1 \otimes \mathcal{A}_2$ by

$$\pi(E) = \sum_{j=1}^n \mu(A_j)\nu(B_j)$$

where $E \in \mathcal{A}_1 \otimes \mathcal{A}_2$ is a finite disjoint union of the rectangles $A_j \times B_j$, $j = 1, \ldots, n$. The *product measure* $\mu \times \nu$ of μ and ν is defined as the extension of the premeasure π to $\mathcal{A}_1 \otimes \mathcal{A}_2$.

Proposition A.23. • *The collection of finite disjoint unions of rectangles is an algebra generating $\mathcal{A}_1 \otimes \mathcal{A}_2$.*

- *With σ-finite measures μ, ν also the product measure is σ-finite. In this case for all rectangles $A \times B$ it holds*

$$\mu \times \nu(A \times B) = \mu(A)\nu(B).$$

- *For $E \in \mathcal{A}_1 \times \mathcal{A}_2$ the sections $E_a = \{b \in \mathcal{A}_2 : (a,b) \in E\}$ and $E^b = \{a \in \mathcal{A}_1 : (a,b) \in E\}$ are measurable for all $a \in X_1$, $b \in X_2$: $E_a \in \mathcal{A}_2$, $E^b \in \mathcal{A}_2$.*

Lebesgue spaces

Definition A.24. Let $(X_1, \mathcal{A}_1)$, $(X_2, \mathcal{A}_2)$ be measurable spaces. A function $F : X_1 \to X_2$ is called *measurable*, iff $F^{-1}(E) \subset \mathcal{A}_1$ for all $E \in \mathcal{A}_2$.

Proposition A.25.   • *Sums, products, maxima and minima of two measurable functions are measurable. In particular, $f^+ := \max(f,0)$ and $f^- := \max(-f,0)$ are measurable if f is measurable. Limits of measurable functions are also measurable.*

- *Every continuous function between metric or topological spaces is measurable.*

- *In a measurable space (X,A) for a set $E \in \mathcal{A}$ the characteristic function $\mathbb{1}_E$ defined by*

$$\mathbb{1}_E(x) = \left\{ \begin{array}{ll} 1 & ,x \in E \\ 0 & ,otherwise \end{array} \right. \tag{A.8}$$

is measurable. Hence every simple function $\phi = \sum_{i=1}^n a_i \mathbb{1}_{E_i}$, $a_i \in \mathbb{C}$, is measurable.

- *If f is $\mathcal{A}_1 \otimes \mathcal{A}_2$-measurable the sections f_a and f^b defined by $f_a(b) = f^b(a) = f(a,b)$ are $\mathcal{A}_2$ and $\mathcal{A}_1$-measurable, respectively.*

Theorem A.26. *Given a measurable space $(X, \mathcal{A})$, every measurable function $f : X \to [0, \infty]$ is the pointwise limit of a sequence $\{\phi_i\}_{n \in \mathbb{N}}$ of simple functions with*

$$0 \leq \phi_1 \leq \phi_2 \leq \ldots \leq f$$

and $\phi_i \to f$ uniformly on every set where f is bounded. The analogous result holds for $f : X \to \mathbb{C}$ measurable with a sequence $\{\psi_i\}_{n \in \mathbb{N}}$ where

$$0 \leq |\psi_1| \leq |\psi_2| \leq \ldots \leq |f|.$$

Theorem A.26 allows the step-wise construction of the integral of a complex-valued measurable function.

Definition A.27. Let $(X, \mathcal{A}, \mu)$ be a measure space and L^+ be the space of all measurable functions $f : X \to [0, \infty]$. For a simple function $\phi = \sum_{i=1}^n a_i \mathbb{1}_{E_i} \in L^+$ the *integral with respect to μ* is defined by

$$\int \phi \, d\mu = \sum_{i=1}^n a_i \mu(E_i).$$

In particular we define $\int_A \phi \, d\mu = \int \phi \mathbb{1}_A \, d\mu$ for $A \in \mathcal{A}$. For $f \in L^+$ we define the integral by

$$\int f \, d\mu := \sup \left\{ \int \phi \, d\mu : 0 \leq \phi \leq f, \phi \text{ simple} \right\} \in [0, \infty].$$

For $f : X \to \mathbb{R}$ we define the integral by

$$\int f \, d\mu = \int f^+ \, d\mu - \int f^- \, d\mu.$$

If $\int |f| \, d\mu < \infty$ (or equivalently $\int f^+ \, d\mu$, $\int f^- \, d\mu < \infty$) we call $f : X \to \mathbb{R}$ *integrable*. Finally for $f : X \to \mathbb{C}$ with $\Re f$, $\Im f$ integrable, define the integral of f by

$$\int f \, d\mu = \int \Re f \, d\mu + i \int \Im f \, d\mu.$$

Denote the set of all integrable $f : X \to \mathbb{C}$ by $L^1(\mu)$ with the equivalence relation

$$f = g \text{ in } L^1(\mu) \quad :\Longleftrightarrow \quad f = g \quad \mu\text{-almost everywhere.}$$

Finally we state the Fubini-Tonelli Theorem concerning the integration under product measures.

Theorem A.28 (Fubini-Tonelli Theorem). *Let $(X_1, \mathcal{A}_1, \mu)$, $(X_2, \mathcal{A}_2, \nu)$ be σ-finite measure spaces.*

a) (Tonelli) If $f \in L^+(X_1 \times X_2)$ then $g(a) = \int f_a \, d\nu$ and $h(b) = \int f^b \, d\mu$ are in $L^+(X_1)$ and $L^+(X_2)$, respectively, and

$$\int f \, d(\mu \times \nu) = \int \int f(a,b) \, d\nu(b) \, d\mu(a) = \int \int f(a,b) \, d\mu(a) \, d\nu(b). \qquad (A.9)$$

b) (Fubini) If $f \in L^1(\mu \times \nu)$ then for μ-almost every $a \in X_1$ and ν-almost every $b \in X_2$ it holds that $f_a \in L^1(\nu)$ and $f_b \in L^1(\mu)$. Further on, $g \in L^1(\mu)$, $h \in L^1(\nu)$ for the almost everywhere defined functions $g(a) = \int f_a \, d\nu$, $h(b) = \int f^b \, d\mu$ and (A.9) holds.

Definition A.29. The *Lebesgue measure on $\mathbb{R}^n$* is defined as the product measure $\lambda_n = \lambda \otimes \ldots, \otimes \lambda$ of the Lebesgue measure on the real line. We denote the Lebesgue measurable sets by $\mathcal{L}^n = \mathcal{L} \otimes \ldots, \otimes \mathcal{L}$.

Remark. *The Lebesgue integral of $f \in L^1(\lambda)$ is sometimes denoted by $\int f(x) \, dx = \int f \, d\lambda$.*
The Lebesgue measure λ_Ω on a measurable set $\Omega \in \mathbb{R}^n$ can be defined as the restriction of λ_n in the sense of $\lambda_\Omega(A) = \lambda_n(\Omega \cap A)$ for all measurable sets A. We identify λ_Ω with λ_n if the support is clear. The Lebesgue space corresponding to such a restriction of the Lebesgue measure on a measurable set is denoted by $L^1(\Omega)$. In particular $L^1(\mathbb{R}^n) = L^1(\lambda_n)$.
The Lebesgue measure has some nice properties such as translation invariance and rotation invariance. For arbitrary linear transformations the following theorem holds.

Theorem A.30 (Transformation Theorem). *Let $S \in GL(n, \mathbb{R})$ be given. If $E \in \mathcal{L}^n$ then $S(E) \in \mathcal{L}^n$ and $\lambda(S(E)) = |\det S| \lambda(E)$.*
Moreover, for every Lebesgue measurable function f on $\mathbb{R}^n$ the composition $f \circ S$ is Lebesgue measurable. For $f \in L^1(\lambda)$ it holds

$$\int f(x) \, dx = |\det(S)| \int f \circ S(x) \, dx. \qquad (A.10)$$

Borel measures, Riesz and convergence of measures

For better readability in the following, we restrict ourselves on a separable metric space X even though many constructions can be generalized to more general spaces.

Definition A.31. Let X be a separable metric space. A measure μ on X is called *Borel measure* if μ operates on the Borel sets $\mathcal{B}(X)$ and $\mu(K) < \infty$ for all compact sets $K \in \mathcal{B}(X)$.

Remark. *Obviously, every finite measure on the Borel sets is a Borel measure, in particular any Dirac measure.*
The Lebesgue measure λ_n on $\mathcal{B}(\mathbb{R}^n)$ is by definition a Borel measure.
On separable metric spaces every finite Borel measure μ is regular, i.e. for a Borel set $A \in \mathcal{B}(X)$ it holds

$$\mu(A) = \inf\{\mu(O) : B \subset O, O \ open\} = \sup\{\mu(K) : K \subset B, K \ compact\}.$$

Theorem A.32 (Riesz Representation Theorem). *Let X be a separable metric space. For every positive bounded linear functional $I : C_c(X) \to \mathbb{R}$, there is a unique Borel measure μ_I with*

$$I(\varphi) = \int \varphi \, d\mu_I \quad for \ all \ \varphi \in C_c(X).$$

Now we define spaces of Borel measures and will give structural information about these spaces.

Definition A.33. For a separable metric space X define

$$\mathcal{M}(X) := \{\mu : \mathcal{B}(X) \to [0,\infty] : \mu \text{ is a Borel measure}\},$$
$$\mathcal{M}^f(X) := \{\mu \in \mathcal{M}(X) : \mu(X) < \infty\},$$
$$\mathcal{M}^1(X) := \{\mu \in \mathcal{M}(X) : \mu(X) = 1\}.$$

Remark. • *By definition $\mathcal{M}^1(X) \subset \mathcal{M}^f(X) \subset \mathcal{M}(X)$, $\mathcal{M}(X)$ is a cone, $\mathcal{M}^f(X)$ a subcone and $\mathcal{M}^1(X)$ a simplex.*

• *Heading towards a convergence notion on $\mathcal{M}(X)$ we remark that the obvious choice for $\mu_n \to \mu$ is given by*

$$\lim_{n \to \infty} \mu_n(A) = \mu(A) \quad for \ all \ A \in \mathcal{B}. \tag{A.11}$$

But this type of convergence is not natural since for instance the Dirac measures δ_{x_n} do not converge to δ_x for $x_n \xrightarrow{n \to \infty} x$, choose $A = \{x\}$ in (A.11) for example. We present a more suitable convergence notion in the following.

• *By the Riesz Representation Theorem there is a bijection between $\mathcal{M}(X)$ and*

$$\{I : C_c(X) \to \mathbb{R}_+ : I \text{ is a bounded linear functional on } X\}.$$

This motivates the following definition

Definition A.34. Given $\mu \in \mathcal{M}(X)$ we say that a sequence $(\mu_n)_n$ in $\mathcal{M}(X)$ *converges vaguely to μ ($\mu_n \xrightarrow{v} \mu$), iff*

$$\int f \, d\mu_n \xrightarrow{n \to \infty} \int f \, d\mu \quad \text{for all } f \in C_c(X). \tag{A.12}$$

Remark. *By the Riesz Representation Theorem $\mu_n \xrightarrow{v} \mu$ if and only if for all $f \in C_c(X)$ the limit $\lim(\int f \, d\mu_n) < \infty$ exists. Hence the vague limit is uniquely defined.*

For $\mathcal{M}^f(X)$, the set of test functions can be extended:

Definition A.35. Given $\mu \in \mathcal{M}^f(X)$ we say that a sequence $(\mu_n)_n$ in $\mathcal{M}^f(X)$ *converges weakly to μ ($\mu_n \xrightarrow{w} \mu$), iff*

$$\int f \, d\mu_n \xrightarrow{n \to \infty} \int f \, d\mu \quad \text{for all } f \in C_b(X). \tag{A.13}$$

Remark. *Since we deal with finite measures $\mu \in \mathcal{M}^f(X)$ it holds $C_b(X) \subset L^1(\mu)$ and the definition makes sense. The following implication of the regularity of finite measures $\mu.\nu \in \mathcal{M}^f(X)$ in particular shows uniqueness of the limit:*

$$\int f \, d\mu = \int f \, d\nu \quad \text{for all } f \in C_b(X) \;\Rightarrow\; \mu = \nu. \tag{A.14}$$

Weak convergence is in general a real subclass of vague convergence:

Theorem A.36. *A sequence $(\mu_n)_n$ in $\mathcal{M}^f(X)$ converges weakly to $\mu \in \mathcal{M}^f(X)$ if and only if $(\mu_n)_n$ converges vaguely to μ and $\lim \mu_n(X) = \mu(X)$.*

We can define the weak topology in $\mathcal{M}^f(X)$ corresponding to weak convergence by using fundamental neighborhoods

$$U_{f_1,\ldots,f_n,\varepsilon}(\mu_0) = \{\mu \in \mathcal{M}^f(X) : \big| \int f_j \, d\mu - \int f_j \, d\mu_0 \big| < \varepsilon \quad \text{for all } 1 \le j \le n\}.$$

It is the coarsest topology where the mappings $\mu \mapsto \int f \, d\mu$ are continuous for every $f \in C_b(X)$. The weak topology is Hausdorff by (A.14). The following theorem gives a criterion of when the equality in (A.11) holds:

Theorem A.37 (Portmanteau Theorem). *Let X be a separable metric space, $\mu \in \mathcal{M}^f(X)$ and $(\mu_n)_n$ be a sequence in $\mathcal{M}^f(X)$. Then it is equivalent:*

- $\mu_n \xrightarrow{w} \mu$.

- $\lim \int f \, d\mu_n = \int f \, d\mu \quad$ *for all $f \in C_{b,u}(X) = \{f \in C_b(X) : f \text{ uniformly continuous}\}$.*

- $\lim \mu_n(X) = \lim \mu(X)$ *and* $\limsup\limits_{n \to \infty} \mu_n(A) \le \mu(A)$ *for all $A \in X$ closed.*

- $\lim \mu_n(X) = \lim \mu(X)$ *and* $\liminf\limits_{n \to \infty} \mu_n(O) \ge \mu(O)$ *for all $O \in X$ open.*

- *If $B \in \mathcal{B}(X)$ is μ-continuous, i.e. $\mu(\partial B) = 0$, then $\lim \mu_n(B) = \mu(B)$.*

While for a general separable metric space X the space $\mathcal{M}^1(X)$ can be metrized using the Prohorov metric (see Chapter 4), on a compact metric space we have the following result.

Theorem A.38. *Let X be a compact metric space. Then $(\mathcal{M}^1(X), d_w)$ is a compact metric space with the metric*

$$d_w(\mu, \nu) = \sum_{i=1}^{\infty} 2^{-i} \left| \int f_i \, d\mu - \int f_i \, d\nu \right| \quad \text{for } \mu, \nu \in \mathcal{M}^1(X) \qquad (A.15)$$

where $(f_i)_i$ is a dense sequence in $C(X)$.

Proof. see [Mañ87]. $\qquad\qquad\square$

Remark. *By Theorem 4.2 the metric d_w is equivalent to the Prohorov metric p. The compactness follows again by the Riesz Representation Theorem. The existence of the dense function sequence is given by the Stone-Weierstraß Theorem.*

Higher Lebesgue- and Sobolev spaces

Analogously to $L^1(\mu)$, we define the Lebesgue spaces $L^p(\mu)$ for $p > 1$.

Definition A.39. *Let a measure space $(X, \mathcal{A}, \mu)$ be given. Define for $p \in \mathbb{R}$, $1 \le p < \infty$ the space $L^p(\mu)$ by*

$$L^p(\mu) = \{f : X \to \mathbb{C} : f \text{ is measurable with } |f|^p \in L^1(\mu)\}.$$

For $p = \infty$ we call a measurable function f essentially bounded, iff

$$\sup_{x \in S \setminus N} |f(x)| < \infty \text{ for a } \mu\text{-null set } N \subset \mathcal{A}.$$

Now define

$$L^\infty(\mu) = \{f : X \to \mathbb{C} : f \text{ is measurable and essentially bounded}\}.$$

and supply the spaces L^p, $p \in [1, \infty]$ with the equivalence relation as above.

Proposition A.40. *The following properties hold for L^p-spaces:*

- *For $p \in [0, \infty]$ the space $L^p(\mu)$ is a Banach space with the norm*

$$\|f\|_p = \left(\int |f|^p \, d\mu \right)^{1/p} \quad \text{for } 1 \le p < \infty, \text{ and}$$

$$\|f\|_\infty = \inf_{N \subset S : \mu(N) = 0} \left(\sup_{x \in S \setminus N} |f(x)| \right).$$

- *For $p = 2$ the space $L^2(\mu)$ is a Hilbert space with scalar product*

$$\langle f, g \rangle_2 = \int fg \, d\mu.$$

- *The Hölder inequality holds: For $f \in L^p(\mu)$ and $g \in L^q(\mu)$ with $p, q \in [1, \infty]$, $\frac{1}{p} + \frac{1}{q} = 1$ the product fg is in $L^1(\mu)$ with*

$$\|fg\|_1 \leq \|f\|_p \|g\|_q$$

For the Lebesgue measure we define Sobolev spaces in the common way by introducing weak derivatives for L^p functions. Therefore define $f^{(\alpha)}$ for a given $f \in L^p(\Omega)$ on a open set $\Omega \subset \mathbb{R}^n$ by

$$\int \partial^\alpha \varphi f \, d\lambda_n = (-1)^{|\alpha|} \int \varphi f^{(\alpha)} \, d\lambda_n \quad \text{for all } \varphi \in C_c^\infty(\Omega). \tag{A.16}$$

Definition A.41. On an open set $\Omega \subset \mathbb{R}^n$ define the *Sobolev space* to the order m and exponent p by

$$W^{m,p}(\Omega) = \{f \in L^p(\Omega) : f^{(\alpha)} \in L^p(\Omega) \text{ defined by (A.16) exists for all } |\alpha| \leq m\}.$$

Define a norm on $W^{m,k}(\Omega)$ by

$$\|f\|_{m,p} = \sum_{|\alpha| \leq m} \|f^{(\alpha)}\|_p.$$

Remark.   • *For $m \in \mathbb{N}$ and $p \in [1, \infty]$ the spaces $W^{m,k}(\Omega)$ are Banach spaces.*

- *For $1 \leq p \leq \infty$ the space of $C^\infty(\Omega)$ functions is dense in $W^{m,p}(\Omega)$, $m \in \mathbb{N}$.*

- *As for Lebesgue spaces important indices are given by $p = 2$, $m \in \mathbb{N}$: the spaces $W^{m,2}(\Omega)$ are Hilbert spaces with the obvious inner product. We denote them by*

$$H^m(\Omega) := W^{m,2}(\Omega). \tag{A.17}$$

- *We denote the weak derivatives also by $\partial^\alpha f = f^{(\alpha)}$ if the context is clear.*

We define the spaces $W_0^{m,p}(\Omega)$ of Sobolev functions satisfying boundary conditions.

Definition A.42. On an open set $\Omega \subset \mathbb{R}^n$ and for indices $m \in \mathbb{N}$, $1 \leq p < \infty$ we define

$$W_0^{m,p}(\Omega) := \{f \in W^{m,p}(\Omega) : \text{ There exist } f_k \in C_c^\infty(\Omega) \text{ with } \|f - f_k\|_{m,p} \xrightarrow{k \to \infty} 0\}.$$

Remark.

$W_0^{m,p}(\Omega)$ is a closed subspace of $W^{m,p}(\Omega)$, hence a Banach space and a Hilbert space for $p = 2$. Accordingly we denote them by

$$H_0^m(\Omega) := W_0^{m,2}(\Omega).$$

Via trace theorems one can define operators that assign boundary values to Sobolev functions in a proper way. By that e.g. Dirichlet boundary conditions can be formulated for Sobolev functions.

A.5   Topics from Fourier analysis

We give a brief survey of the Fourier analysis for functions and distributions in $\mathbb{R}^n$ used in Section 4.1.4. For more details and proofs of the following statements see for instance [Fol99] or [Gra08].

Definition A.43. Define the *Schwartz space* $\mathcal{S} = \mathcal{S}(\mathbb{R}^n) \subset C^\infty$ by

$$\mathcal{S} = \{f \in C^\infty : \|f\|_{(N,\alpha)} < \infty \text{ for all } N \in \mathbb{N}, \text{ and multi-indices } \alpha\} \tag{A.18}$$

where $\|f\|_{(N,\alpha)} = \sup\limits_{x \in \mathbb{R}^n} \left(1 + |x|^N\right) |\partial^\alpha f(x)|$.

Remark. *By definition,* $C_c^\infty \subset \mathcal{S} \subset C_0$.
$C_c(\mathbb{R}^n)$ *and* $\mathcal{S}(\mathbb{R}^n)$ *are nonempty, e.g.* $\psi \in C_c^\infty$ *for* $\psi(x) = e^{|x|^2 - 1}$ *if* $|x| < 1$, $\psi(x) = 0$, $|x| \geq 1$. *Further,* C_c^∞ *(and hence also* $\mathcal{S}$*) is dense in* $L^p(\mathbb{R}^n)$, $1 \leq p < \infty$, *and in* $C_0(\mathbb{R}^n)$.
If $f \in \mathcal{S}(\mathbb{R}^n)$, $g \in \mathcal{S}(\mathbb{R}^m)$ *we have* $(x, y) \mapsto f(x)g(y) \in \mathcal{S}(\mathbb{R}^{n+m})$.

Definition A.44. Define the *Fourier transform* of $f \in L^1(\mathbb{R}^m)$ by

$$[\mathcal{F}(f)](\xi) = \widehat{f}(\xi) = \int e^{-2\pi \xi^T x} f(x) \, dx.$$

Proposition A.45. *We collect some properties of* $\widehat{f}$ *for* $f \in L^1(\mathbb{R}^n)$:

a) $[\mathcal{F}(\tau_y f)](\xi) = e^{-2\pi i \xi^T y} \widehat{f}(\xi)$ *and* $\tau_\eta(\widehat{f}) = \widehat{h}$ *with* $h(x) = e^{-2\pi i \eta^t x} f(x)$ *where* $\tau_y f(x) = f(x - y)$.

b) $\mathcal{F}(f \circ S) = |\det S|^{-1} \widehat{f} \circ (S^T)^{-1}$ *for every* $S \in GL(n, \mathbb{R})$.

c) *By definition* $\mathcal{F}(L^1(\mathbb{R}^n)) \subset C_0(\mathbb{R}^n)$.

d) *If* $f \in C^k$, $\partial^\alpha f \in L^1$ *for* $|\alpha| \leq k$ *and* $\partial^\alpha f \in C_0$ *for* $|\alpha| \leq k - 1$ *then* $[\mathcal{F}(\partial^\alpha f)](\xi) = (2\pi i \xi)^\alpha \widehat{f}(\xi)$.

e) $\mathcal{F} : \mathcal{S} \to \mathcal{S}$ *is continuous. Moreover* $\mathcal{F}$ *is an isomorphism of* $\mathcal{S}$ *onto itself.*

f) *For* $f, g \in L^1$ *it holds* $\int \widehat{f} g \, d\lambda_n = \int f \widehat{g} \, d\lambda_n$.

g) *If* $f \in \mathcal{S}(\mathbb{R}^n)$ *and* $g \in \mathcal{S}(\mathbb{R}^m)$ *then the Fourier transform of* $h : (x, y) \mapsto f(x)g(y) \in \mathcal{S}(\mathbb{R}^{n+m})$ *is given by* $\widehat{h}(\xi_1, \xi_2) = \widehat{f}(\xi_1)\widehat{g}(\xi_2)$.

Remark. *For the proof of Theorem A.45 e) the inverse of the Fourier transform defined by*

$$f^\vee(x) = \widehat{f}(-x) = \int e^{2\pi i \xi^T x} f(\xi) \, d\xi$$

is used. For $\widehat{f}, f^\vee \in L^1$ *it holds* $(\widehat{f})^\vee = (f^\vee)^\wedge = f_0$ *a.e. where* $f_0 \in C(\mathbb{R}^n)$.

Theorem A.46 (Plancherel). *If* $f \in L^1 \cap L^2$ *then* $\widehat{f} \in L^2$. $\mathcal{F}_{|L^1 \cap L^2}$ *extends uniquely to a unitary isomorphism on* L^2.

Definition A.47. Define the set of distributions on $\mathbb{R}^n$ by $\mathcal{D}' = \mathcal{D}'(\mathbb{R}^n)$ with

$$\mathcal{D}' = \{F : C_c^\infty(\mathbb{R}^n) \to \mathbb{R} \text{ linear and bounded}\}.$$

We denote the value of $F \in \mathcal{D}'$ at $\phi \in C_c^\infty(\mathbb{R}^n)$ by $\langle F, \phi \rangle$ to avoid notational confusion with the following identifications.

Remark. *A function $f \in L^1(\mathbb{R}^n)$ defines a distribution via $\phi \mapsto \int f\phi\, d\lambda$. In particular $L^1(\mathbb{R}^n) \subset \mathcal{D}'$.*
A Borel measure $\mu \in \mathcal{M}(Q)$ with $Q \subset \mathbb{R}^n$ compact defines a distribution via $\phi \mapsto \int \phi\, d\mu$. We identify $f \in L^1$ and $\mu \in \mathcal{M}(Q)$ with the distribution defined above.
With this notation in particular the Dirac measure $\delta_0 = \delta_0^n$ at $0 \in \mathbb{R}^n$ given by (A.7) is a distribution with $\langle \delta_0, \phi \rangle = \phi(0)$.

The following definition deals with extensions of linear operations from functions to distributions:

Definition A.48. Let T be a linear map from $L^1(\mathbb{R}^n)$ onto itself and T' a corresponding linear operator on $C_c^\infty(\mathbb{R}^n)$ with

$$\int (Tf)\phi\, d\lambda_n = \int f(T'\phi)\, d\lambda_n \quad \text{for all } f \in L^1(\mathbb{R}^n),\ \phi \in C_c^\infty(\mathbb{R}^n).$$

Then we define the linear operator in $\mathcal{D}'$ associated to T and identified with T by

$$\langle Tf, \phi \rangle = \langle F, T'\phi \rangle \quad \text{for all } F \in \mathcal{D}',\ \phi \in C_c^\infty(\mathbb{R}^n). \tag{A.19}$$

Remark. *Examples are given by the translation $T = \tau_y$ where $\langle \tau_y F, \phi \rangle = \langle F, \tau_{-y}\phi \rangle$ and the linear transformation $Tf = f \circ S$ given by the matrix $S \in GL(n, \mathbb{R})$ with corresponding operator $T'\phi = |\det S|^{-1}\phi \circ S^{-1}$. Then (A.19) implies*

$$\langle F \circ S, \phi \rangle = |\det S|^{-1}\langle F, \phi \circ S^{-1} \rangle \tag{A.20}$$

Definition A.49. A continuous linear functional on the Schwartz space $\mathcal{S}$ is called a *tempered distribution*. As usual, we denote the space of tempered distributions by $\mathcal{S}'$.

Proposition A.50. *Every compactly supported distribution is tempered. In particular $\mathcal{M}(Q) \subset \mathcal{S}'$ for compact $Q \subset \mathbb{R}^n$. For $\psi \in C_c^\infty$ and a tempered distribution $F \in \mathcal{S}'$ the product ψF is a tempered distribution iff ψ is slowly increasing, i.e. for every multi-index α there is an $N = N(\alpha) \in \mathbb{N}$ with*

$$|\partial^\alpha \psi(x)| \le C_\alpha (1 + |x|)^N.$$

Definition A.51. For tempered distributions define the *Fourier transform* $\mathcal{F} : \mathcal{S}' \to \mathcal{S}'$ by

$$\langle \widehat{F}, \phi \rangle = \langle F, \widehat{\phi} \rangle \quad \text{for all } F \in \mathcal{S}',\ \phi \in \mathcal{S}.$$

Remark. *By Theorem A.45 e) and f) the Fourier transform for tempered distributions is well-defined and it clearly agrees with the definition on $L^1(\mathbb{R}^n)$.*
Most properties of $\mathcal{F}$ on L^1 continue to hold for $\mathcal{F} : \mathcal{S}' \to \mathcal{S}'$. In particular

$$\mathcal{F}(\tau_y F) = e^{-2\pi \xi^T y}\widehat{F},\ \ \tau_\eta \widehat{F} = \mathcal{F}(e^{2\pi i \eta^T x}F) \quad and \tag{A.21}$$

$$\mathcal{F}(F \circ S) = |\det S|^{-1}\widehat{F} \circ (S^T)^{-1} \quad for\ S \in GL(n, \mathbb{R}). \tag{A.22}$$

For products of measures μ, ν with compact support the Fourier transform of the product measure $\mu \times \nu$ defined by Definition A.22 is given by

$$\mathcal{F}(\mu \times \nu) = \mathcal{F}(\mu)\mathcal{F}(\nu). \tag{A.23}$$

Using $\mathrm{sinc}(x) = \frac{\sin(\pi x)}{\pi x}$, $x \neq 0$, $\mathrm{sinc}(0) = 1$ *we state two basic Fourier transforms*

$$\widehat{\delta_0} = 1, \qquad \mathcal{F}\big(\mathbb{1}_{[-\frac{1}{2},\frac{1}{2}]}\big) = \mathrm{sinc}. \tag{A.24}$$

As before, the inverse transform is defined by

$$\langle F^\wedge, \phi \rangle = \langle F, \phi^\wedge \rangle.$$

For $F \in \mathcal{S}'$ it holds:

$$F = (\widehat{F})^\wedge = (F^\wedge)^\vee.$$

Hence $\mathcal{F}$ is an isomorphism on $\mathcal{S}'$.

Negative Sobolev norms

We derive a generalization of the Sobolev spaces H_k, $k \in \mathbb{N}$, as defined in (A.17) to spaces H_s, $s \in \mathbb{R}$ in terms of the Fourier transformation. Therefore observe that Theorem A.46 and Theorem A.45 d) imply that $f \in H_k$ iff $\xi \mapsto \xi^\alpha \widehat{f}(\xi) \in L^2$ for $|\alpha| \leq k$. Further, a simple calculation shows the existence of $C_1, C_2 > 0$ with

$$C_1(1 + |\xi|^2)^k \leq \sum_{|\alpha| \leq k} |\xi^\alpha|^2 \leq C_2(1 + |\xi|^2)^k$$

for all $\xi \in \mathbb{C}$. This implies $f \in H_k$ iff $(1 + |\xi|^2)^{k/2}\widehat{f} \in L^2$ and equivalence of the norms $\| \cdot \|_{2,k}$ and $f \mapsto \|(1 + |\xi|^2)^{k/2}\widehat{f}\|_2$.

Definition A.52. For $s \in \mathbb{R}$ define the Sobolev space $H_s = H_s(\mathbb{R}^n)$ by

$$H_s = \{f \in \mathcal{S}' : \Lambda_s f \in L^2\}$$

where $\Lambda_s f = \left[(1 + |\xi|^2)^{s/2}\widehat{f}\right]^\vee$. Define an inner product and a norm on H_s by

$$\langle f, g \rangle_{(s)} = \langle \Lambda_s f, \Lambda_s g \rangle_2 = \int \widehat{f}(1 + |\xi|^2)^s \overline{\widehat{g}}\, d\xi,$$

$$\|f\|_{(s)} = \|\Lambda_s f\|_2 = \left[\int |\widehat{f}(\xi)|^2(1 + |\xi|^2)^s\, d\xi\right]^{1/2}.$$

Remark. • *Recall that by Theorem A.46 $\|f\|_2 = \|f^\wedge\|_2 = \|\widehat{f}\|_2$ for $f \in L^2$. Further on $(1 + |\cdot|^2)^{s/2}$ is a slowly increasing function in the sense of Theorem A.50. It follows that Λ_s is an isomorphism with $\Lambda_s^{-1} = \Lambda_{-s}$.*

• *H_s is indeed a Hilbert space for all $s \in \mathbb{R}$.*

• *For all $s \in \mathbb{R}$ it holds $\mathcal{S} \subset H_s$.*

• *$H_0 = L^2$ and Definition A.52 is a proper generalization of the previous definition of Sobolev spaces, i.e. it agrees with the Definition (A.17) for $k \in \mathbb{N}$.*

• *The Dirac measure $\delta_0 \in \mathcal{S}'(\mathbb{R}^n)$ is in H_s iff $s < \frac{1}{2}n$.*

List of Figures

Bibliography

[Alt99]   H. W. Alt. *Lineare Funktionalanalysis*. Springer-Verlag, Berlin, third edition, 1999.

[Ant05]   A. C. Antoulas. *Approximation of large-scale dynamical systems, Advances in Design and Control*, volume 6. Society for Industrial and Applied Mathematics (SIAM), Philadelphia, PA, 2005.

[AR87]   K. M. Anstreicher and U. G. Rothblum. Using Gauss-Jordan elimination to compute the index, generalized nullspaces, and Drazin inverse. *Linear Algebra Appl.*, 85, pp. 221–239, 1987.

[ASG01]   A. C. Antoulas, D. C. Sorensen, and S. Gugercin. A survey of model reduction methods for large-scale systems. In *Structured matrices in mathematics, computer science, and engineering, I (Boulder, CO, 1999), Contemp. Math.*, volume 280, pp. 193–219. Amer. Math. Soc., 2001.

[Bal00]   V. Baladi. *Positive transfer operators and decay of correlations, Advanced Series in Nonlinear Dynamics*, volume 16. World Scientific Publishing Co. Inc., River Edge, NJ, 2000.

[Bau92]   H. Bauer. *Maß- und Integrationstheorie*. de Gruyter Lehrbuch. Walter de Gruyter & Co., Berlin, second edition, 1992.

[BBP95]   J. Bastero, J. Bernués, and A. Peña. An extension of Milman's reverse Brunn-Minkowski inequality. *Geom. Funct. Anal.*, 5(3), pp. 572–581, 1995.

[BIG03]   A. Ben-Israel and T. N. E. Greville. *Generalized inverses*. CMS Books in Mathematics/Ouvrages de Mathématiques de la SMC, 15. Springer-Verlag, New York, second edition, 2003.

[Bil99]   P. Billingsley. *Convergence of probability measures*. Wiley Series in Probability and Statistics: Probability and Statistics. John Wiley & Sons Inc., New York, second edition, 1999.

[Bir57]   G. Birkhoff. Extensions of Jentzsch's theorem. *Trans. Amer. Math. Soc.*, 85, pp. 219–227, 1957.

[BKL02]   M. Blank, G. Keller, and C. Liverani. Ruelle-Perron-Frobenius spectrum for Anosov maps. *Nonlinearity*, 15(6), pp. 1905–1973, 2002.

[BP79]   A. Berman and R. J. Plemmons. *Nonnegative matrices in the mathematical sciences.* Academic Press [Harcourt Brace Jovanovich Publishers], New York, 1979.

[BV92]   A. V. Babin and M. I. Vishik. *Attractors of evolution equations, Studies in Mathematics and its Applications,* volume 25. North-Holland Publishing Co., Amsterdam, 1992.

[CKR08]   T. Caraballo, P. E. Kloeden, and J. Real. Invariant measures and statistical solutions of the globally modified Navier-Stokes equations. *Discrete Contin. Dyn. Syst. Ser. B,* 10(4), pp. 761–781, 2008.

[Cli68]   R. E. Cline. Inverses of rank invariant powers of a matrix. Summary of results. In *Proc. Sympos. Theory and Application of Generalized Inverses of Matrices (Lubbock, Texas, 1968),* pp. 47–52. Texas Tech. Press, Lubbock, Tex., 1968.

[CM79]   S. L. Campbell and C. D. Meyer, Jr. *Generalized inverses of linear transformations, Surveys and Reference Works in Mathematics,* volume 4. Pitman (Advanced Publishing Program), Boston, Mass., 1979.

[DDL93]   J. Ding, Q. Du, and T. Y. Li. High order approximation of the Frobenius-Perron operator. *Appl. Math. Comput.,* 53(2-3), pp. 151–171, 1993.

[DFJ01]   M. Dellnitz, G. Froyland, and O. Junge. The algorithms behind GAIO-set oriented numerical methods for dynamical systems. In *Ergodic theory, analysis, and efficient simulation of dynamical systems,* pp. 145–174, 805–807. Springer-Verlag, Berlin, 2001.

[DHJR97]   M. Dellnitz, A. Hohmann, O. Junge, and M. Rumpf. Exploring invariant sets and invariant measures. *Chaos,* 7(2), pp. 221–228, 1997.

[DJ98]   M. Dellnitz and O. Junge. An adaptive subdivision technique for the approximation of attractors and invariant measures. *Comput. Visual. Sci.,* 1, pp. 63–68, 1998.

[DJ99]   M. Dellnitz and O. Junge. On the approximation of complicated dynamical behavior. *SIAM J. Numer. Anal.,* 36(2), pp. 491–515, 1999.

[DJ02]   M. Dellnitz and O. Junge. Set oriented numerical methods for dynamical systems. In *Handbook of dynamical systems,* volume 2, pp. 221–264. North-Holland, Amsterdam, 2002.

[DKP95]   P. Diamond, P. Kloeden, and A. Pokrovskii. Interval stochastic matrices: a combinatorial lemma and the computation of invariant measures of dynamical systems. *J. Dynam. Differential Equations,* 7(2), pp. 341–364, 1995.

[Doo67]   J. L. Doob. *Stochastic processes.* John Wiley & Sons Inc., New York, seventh edition, 1967.

[Dud02]   R. M. Dudley. *Real analysis and probability, Cambridge Studies in Advanced Mathematics,* volume 74. Cambridge University Press, Cambridge, 2002.

[DZ96]   J. Ding and A. Zhou. Finite approximations of Frobenius-Perron operators. A solution of Ulam's conjecture to multi-dimensional transformations. *Phys. D*, 92(1-2), pp. 61–68, 1996.

[Els05]   J. Elstrodt. *Maß- und Integrationstheorie*. Springer Textbook. Springer-Verlag, Berlin, fourth edition, 2005.

[FM86]   R. E. Funderlic and C. D. Meyer, Jr. Sensitivity of the stationary distribution vector for an ergodic Markov chain. *Linear Algebra Appl.*, 76, pp. 1–17, 1986.

[Fol99]   G. B. Folland. *Real analysis*. Pure and Applied Mathematics. John Wiley & Sons Inc., New York, second edition, 1999. Modern techniques and their applications, A Wiley-Interscience Publication.

[Fro95]   G. Froyland. Finite approximation of Sinai-Bowen-Ruelle measures for Anosov systems in two dimensions. *Random Comput. Dynam.*, 3(4), pp. 251–263, 1995.

[Gar02]   R. J. Gardner. The Brunn-Minkowski inequality. *Bull. Amer. Math. Soc. (N.S.)*, 39(3), pp. 355–405, 2002.

[Gau62]   W. Gautschi. On inverses of Vandermonde and confluent Vandermonde matrices. *Numer. Math.*, 4, pp. 117–123, 1962.

[GK09]   M. Griebel and S. Knapek. Optimized general sparse grid approximation spaces for operator equations. *Math. Comp.*, 78(268), pp. 2223–2257, 2009.

[Gra08]   L. Grafakos. *Classical Fourier analysis, Graduate Texts in Mathematics*, volume 249. Springer-Verlag, New York, second edition, 2008.

[GvL96]   G. H. Golub and C. F. van Loan. *Matrix computations*. Johns Hopkins Studies in the Mathematical Sciences. Johns Hopkins University Press, Baltimore, third edition, 1996.

[Hal88]   J. K. Hale. *Asymptotic behavior of dissipative systems, Mathematical Surveys and Monographs*, volume 25. American Mathematical Society, Providence, RI, 1988.

[Har81]   R. E. Hartwig. A method for calculating A^d. *Math. Japon.*, 26(1), pp. 37–43, 1981.

[Hen81]   D. Henry. *Geometric theory of semilinear parabolic equations, Lecture Notes in Mathematics*, volume 840. Springer-Verlag, Berlin, 1981.

[HJ85]   R. A. Horn and C. R. Johnson. *Matrix analysis*. Cambridge University Press, Cambridge, 1985.

[HJ94]   R. A. Horn and C. R. Johnson. *Topics in matrix analysis*. Cambridge University Press, Cambridge, 1994.

[HLB96]   P. Holmes, J. L. Lumley, and G. Berkooz. *Turbulence, coherent structures, dynamical systems and symmetry*. Cambridge Monographs on Mechanics. Cambridge University Press, Cambridge, 1996.

[HNR90]　　R. E. Hartwig, M. Neumann, and N. J. Rose. An algebraic-analytic approach to nonnegative bases. *Linear Algebra Appl.*, 133, pp. 77–88, 1990.

[IK03]　　　P. Imkeller and P. Kloeden. On the computation of invariant measures in random dynamical systems. *Stoch. Dyn.*, 3(2), pp. 247–265, 2003.

[Ise09]　　　A. Iserles. *A first course in the numerical analysis of differential equations.* Cambridge Texts in Applied Mathematics. Cambridge University Press, Cambridge, second edition, 2009.

[Jun01]　　　O. Junge. Almost invariant sets in chua's circuit. *Int. J. Bif. and Chaos*, 7, pp. 2475–2485, 2001.

[Jun09]　　　O. Junge. Discretization of the Frobenius-Perron operator using a sparse Haar tensor basis—the Sparse Ulam method. *SIAM J. Num. Anal.*, 46(5), pp. 3464–3485, 2009.

[Kat76]　　　T. Kato. *Perturbation theory for linear operators.* Springer-Verlag, Berlin, second edition, 1976.

[Kel82]　　　G. Keller. Stochastic stability in some chaotic dynamical systems. *Monatsh. Math.*, 94(4), pp. 313–333, 1982.

[Kem02]　　J. Kemper. *Attraktoren und invariante Maße in Reaktions-Diffusions-Gleichungen.* Master's thesis, Universität Bielefeld, 2002.

[Kem08]　　J. Kemper. Computing invariant measures with dimension reduction methods, 2008. Preprint, available for download at www.math.uni-bielefeld.de/sfb701/cgi-bin/preprints.pl.

[KH95]　　　A. Katok and B. Hasselblatt. *Introduction to the modern theory of dynamical systems, Encyclopedia of Mathematics and its Applications*, volume 54. Cambridge University Press, Cambridge, 1995.

[Kif86]　　　Y. Kifer. General random perturbations of hyperbolic and expanding transformations. *J. Analyse Math.*, 47, pp. 111–150, 1986.

[Kir96]　　　A. Kirsch. *An introduction to the mathematical theory of inverse problems, Applied Mathematical Sciences*, volume 120. Springer-Verlag, New York, 1996.

[KP82]　　　E. Kohlberg and J. W. Pratt. The contraction mapping approach to the Perron-Frobenius theory: why Hilbert's metric? *Math. Oper. Res.*, 7(2), pp. 198–210, 1982.

[Kra86]　　　U. Krause. A nonlinear extension of the Birkhoff-Jentzsch theorem. *J. Math. Anal. Appl.*, 114(2), pp. 552–568, 1986.

[Kra01]　　　U. Krause. Concave Perron-Frobenius theory and applications. *Nonlinear Anal.*, 47(3), pp. 1457–1466, 2001.

[KV01]　　　K. Kunisch and S. Volkwein. Galerkin proper orthogonal decomposition methods for parabolic problems. *Numer. Math.*, 90(1), pp. 117–148, 2001.

[KV02]   K. Kunisch and S. Volkwein. Galerkin proper orthogonal decomposition methods for a general equation in fluid dynamics. *SIAM J. Numer. Anal.*, 40(2), pp. 492–515, 2002.

[Lar99]   S. Larsson. Numerical analysis of semilinear parabolic problems. In *The graduate student's guide to numerical analysis '98 (Leicester), Springer Ser. Comput. Math.*, volume 26, pp. 83–117. Springer-Verlag, Berlin, 1999.

[Li76]   T. Y. Li. Finite approximation for the Frobenius-Perron operator. A solution to Ulam's conjecture. *J. Approximation Theory*, 17(2), pp. 177–186, 1976.

[Liv03]   C. Liverani. Invariant measures and their properties. A functional analytic point of view. In *Dynamical systems. Part II*, Publ. Cent. Ric. Mat. Ennio Giorgi, pp. 185–237. Scuola Norm. Sup., Pisa, 2003.

[Lor63]   E. Lorenz. Deterministic nonperiodic flow. *J. Atmos. Sci*, 20, pp. 130–141, 1963.

[Lou89]   A. K. Louis. *Inverse und schlecht gestellte Probleme.* Teubner Studienbücher Mathematik. [Teubner Mathematical Textbooks]. B. G. Teubner, Stuttgart, 1989.

[LSS94]   S. Larsson and J. M. Sanz-Serna. The behavior of finite element solutions of semilinear parabolic problems near stationary points. *SIAM J. Numer. Anal.*, 31(4), pp. 1000–1018, 1994.

[LT03]   S. Larsson and V. Thomée. *Partial differential equations with numerical methods, Texts in Applied Mathematics*, volume 45. Springer-Verlag, Berlin, 2003.

[LW01]   X. Li and Y. Wei. An improvement on the perturbation of the group inverse and oblique projection. *Linear Algebra Appl.*, 338, pp. 53–66, 2001.

[Mañ87]   R. Mañé. *Ergodic theory and differentiable dynamics.* Springer-Verlag, Berlin, 1987.

[Mey80]   C. D. Meyer, Jr. The condition of a finite Markov chain and perturbation bounds for the limiting probabilities. *SIAM J. Algebraic Discrete Methods*, 1(3), pp. 273–283, 1980.

[Mey94]   C. D. Meyer. Sensitivity of the stationary distribution of a Markov chain. *SIAM J. Matrix Anal. Appl.*, 15(3), pp. 715–728, 1994.

[MHvMD06]   P. G. Mehta, M. Hessel-von Molo, and M. Dellnitz. Symmetry of attractors and the Perron-Frobenius operator. *J. Difference Equ. Appl.*, 12(11), pp. 1147–1178, 2006.

[Min88]   H. Minc. *Nonnegative matrices.* Wiley-Interscience Series in Discrete Mathematics and Optimization. John Wiley & Sons Inc., New York, 1988.

[MY02]   N. Masmoudi and L.-S. Young. Ergodic theory of infinite dimensional systems with applications to dissipative parabolic PDEs. *Comm. Math. Phys.*, 227(3), pp. 461–481, 2002.

[NVL99]   R. D. Nussbaum and S. M. Verduyn Lunel. Generalizations of the Perron-Frobenius theorem for nonlinear maps. *Mem. Amer. Math. Soc.*, 138(659), 1999.

[Osb75]   J. E. Osborn. Spectral approximation for compact operators. *Math. Comput.*, 29, pp. 712–725, 1975.

[Ott09]   D. Otten. *Attraktoren für Finite-Elemente Diskretisierungen parabolischer Differentialgleichungen.* Master's thesis, Universität Bielefeld, 2009.

[Pap05]   A. Papadopoulos. *Metric spaces, convexity and nonpositive curvature, IRMA Lectures in Mathematics and Theoretical Physics*, volume 6. European Mathematical Society (EMS), Zürich, 2005.

[Pil99]   S. Y. Pilyugin. *Shadowing in dynamical systems, Lecture Notes in Mathematics*, volume 1706. Springer-Verlag, Berlin, 1999.

[RCM04]   C. Rowley, T. Colonius, and R. Murray. Model reduction for compressible flows using POD and Galerkin projection. *Physica D Nonlinear Phenomena*, 189(1-2), pp. 115–129, 2004.

[Rob01]   J. C. Robinson. *Infinite-dimensional dynamical systems.* Cambridge Texts in Applied Mathematics. Cambridge University Press, Cambridge, 2001.

[Rue76]   D. Ruelle. A measure associated with Axiom-A attractors. *Amer. J. Math.*, 98(3), pp. 619–654, 1976.

[SH96]   A. M. Stuart and A. R. Humphries. *Dynamical systems and numerical analysis, Cambridge Monographs on Applied and Computational Mathematics*, volume 2. Cambridge University Press, Cambridge, 1996.

[Sin94]   Y. G. Sinaĭ. *Topics in ergodic theory, Princeton Mathematical Series*, volume 44. Princeton University Press, Princeton, NJ, 1994.

[Sma67]   S. Smale. Differentiable dynamical systems. *Bull. Amer. Math. Soc.*, 73, pp. 747–817, 1967.

[Ste73]   G. W. Stewart. *Introduction to matrix computations.* Academic Press, New York-London, 1973.

[Ste90]   G. W. Stewart. *Matrix Perturbation Theory.* Academic Press, New York-London, 1990.

[Tem97]   R. Temam. *Infinite-dimensional dynamical systems in mechanics and physics, Applied Mathematical Sciences*, volume 68. Springer-Verlag, New York, second edition, 1997.

[Ula60]   S. M. Ulam. *A collection of mathematical problems.* Interscience Tracts in Pure and Applied Mathematics, no. 8. Interscience Publishers, New York-London, 1960.

[Wig03]   S. Wiggins. *Introduction to applied nonlinear dynamical systems and chaos*, *Texts in Applied Mathematics*, volume 2. Springer-Verlag, New York, second edition, 2003.

[You02]   L.-S. Young. What are SRB measures, and which dynamical systems have them? *J. Statist. Phys.*, 108(5-6), pp. 733–754, 2002. Dedicated to David Ruelle and Yasha Sinai on the occasion of their 65th birthdays.

Index